UNDER THE WEATHER

JAMES RENWICK

UNDER THE WEATHER

A Future Forecast for New Zealand

HarperCollins*Publishers*

HarperCollins*Publishers*
Australia • Brazil • Canada • France • Germany • Holland • India
Italy • Japan • Mexico • New Zealand • Poland • Spain • Sweden
Switzerland • United Kingdom • United States of America

First published in 2023
by HarperCollins*Publishers* (New Zealand) Limited
Unit D1, 63 Apollo Drive, Rosedale, Auckland 0632, New Zealand
harpercollins.co.nz

A catalogue record for this book is available from the National Library of New Zealand

ISBN 978 1 7755 4172 1 (pbk)
ISBN 978 1 7754 9203 0 (ebook)

Cover design by Megan van Staden
Cover image © Rob Suisted, Nature's Pic Images
Typeset in Minion Pro by Kirby Jones
Printed and bound in Australia by McPherson's Printing Group

FSC
www.fsc.org
MIX
Paper | Supporting
responsible forestry
FSC® C001695

To Muriel and Arthur

Contents

Introduction

I grew up beneath the wide-open skies of central Canterbury, a place exposed to every kind of weather. Wild thunderstorms would sometimes roll overhead, spitting hailstones on the roof and sending gusts of wind to worry at the doors and windows. In winter, I often woke to frosts that solidified every puddle and blade of grass. One crisp and sunny winter's morning, when I was about four, I watched my mum hang out the sheets, and they froze before she'd finished pegging out. I went and grabbed a corner of one sheet and swung it; it was completely solid, a great white board fanning back and forth. That really amazed and intrigued me. How could something normally so soft and comforting become so solid and cold?

Sometimes in winter, we got snow, occasionally so much that school was closed for the day. And in spring and summer, the nor'wester would visit with its friend, the baking heat.

'Conversation about the weather', Oscar Wilde famously said, 'is the last refuge of the unimaginative.' But, the way I see it, the weather is so imaginative it can be difficult *not* to talk about it. That's especially true here in Aotearoa New Zealand. Our long and narrow country, with its northern tip very nearly entering the tropics and its southern tip within waving distance of Antarctica, features something of a Goldilocks climate – it's not too hot and not too cold, although it can be pretty wet and definitely windy. Our dynamic weather and climate are really important for a lot of our economy, and for how we all live our lives here. To those of us who are interested in such things, Aotearoa is the perfect site for studying the interplay of forces that drive our weather, the things that make up our climate system.

I suppose the changeable and occasionally extreme Canterbury weather must have shaped my thinking more than I knew because, after finishing a mathematics degree, I went and trained to be a meteorologist – a weather man. I've had my head in the clouds ever since. It's up there – where massive, drifting, unruly forces whirl about in an ongoing dance of fast and slow, hot and cold – that I've spent much of my career. Sometimes, I come back down to Earth too because, in order to understand the climate, we also need to look deep in the ground beneath our feet, and plumb the ocean's depths.

I've now spent over four decades studying the atmosphere and the climate. I look at the clouds, I pay attention to the winds, I watch the way the air moves, what drives the jet streams, and what controls the endless sequence of ups and downs in our daily weather. It's all been fodder for my studies, and the fascination still hasn't worn off – I doubt it ever will. My day job is teaching students about weather and climate at Te Herenga Waka/Victoria University of Wellington, and in 2018 I was honoured to be awarded the Prime Minister's Science Communication Prize.

Over the past 20 years, I've also become increasingly involved in international science collaboration. That started with helping to write the Intergovernmental Panel on Climate Change (IPCC) Assessment Reports – comprehensive documents that provide summaries of the most robust science on what's happening with our climate to the powers-that-be. Around a decade ago, I also took a spot on the Joint Scientific Committee of the World Climate Research Programme (WCRP), an organisation that coordinates much of the underpinning science that the IPCC documents report on. That posting led to a management role with one of the WCRP's main pillars, the CliC (Climate and Cryosphere) project – a great experience, not least because the cryosphere, or the climate system's 'frozen bits', have been a developing interest through my career. In 2018, I became one of the leaders in the New Zealand Antarctic Science Platform, studying how sea ice is changing in the Ross Sea and what that might look like in a warmer world and, most recently, I became one of eight commissioners on New Zealand's He Pou a Rangi – Climate Change Commission.

Over the decades, and especially through my involvement with those IPCC reports and the work of the WCRP, I've been fortunate to gain a really broad understanding of the climate system. I've had the opportunity to collaborate with experts from all over the world, expanding my knowledge of everything climate-related: the deep ocean, the ice sheets, the upper atmosphere, extreme events like floods and droughts, climate history and the role of life on Earth, and more.

At the most basic level, understanding the climate is a matter of understanding how our fluid atmosphere and oceans respond to heating. It's also a matter of understanding the different forces at work: the fast and responsive atmosphere, the slow and ponderous oceans, the busy winds, the warming sun, the critical greenhouse gases, and the flow of energy through it all. Everything has a part to play, and if there's one thing I've learnt from my many years working in this field, it's this: the climate might be immense, but it's also sensitive. Very sensitive. Small adjustments can have enormous consequences.

Right now, the climate is changing. It's changed in the past, but this time is different. This time it's not an asteroid or an ice age or a volcano causing the change. This time, it's us. The science has been very, very clear for over three decades: human emissions of greenhouse gases are warming our climate in increments that might not seem like a lot, but are in fact enough to change everything about the world as we know it.

Our planet's climate is a magical thing. What a marvel it is that the conditions have been just right for us to live here, to thrive, to build human civilisation and enterprise as we know it. Isn't that incredible? Isn't that something precious, something we should all fight to protect and sustain?

*

This is a book about climate change. It's about what has happened, what is happening and what's going to happen. It's also, crucially, about what we can do to stop climate change –

because there is a lot that we can do. Humans are driving global warming, which means that we as a species can also halt it. That's an enormous responsibility, but it's also where our greatest hope lies.

My hope is that this book will teach you how the climate works, so that you can better understand the climate-change forecast, both in the immediate future and the long term. In Part I, I'll give you an overview of Earth's climate, and what drives it, from the distant past to the present day. In Part II, I'll zero in on New Zealand's climate and what makes it special, taking time to also look at what's going on with our climate today. And, because we are not alone in this South Pacific pocket of the globe, I'll also spend a bit of time going over what's changing for our island neighbours.

In Part III, I'll look to the immediate future. Based on the climate change that's already locked in, what's the forecast for New Zealand and the South Pacific? What can we expect to see in the coming years? What are our land and seas going to look like, and what's going to be happening in our skies?

Finally, in Part IV, I'll set my sights – and yours, too – on the horizon. I'll start by painting a picture of the world's best- and worst-case scenarios. What would happen if we stopped emissions tomorrow? What if we didn't stop at all? What if we burned every last bit of fossil fuel we could get our hands on? Then, once you're armed with the knowledge of our role in how the future will take shape, I'll run through everything we need to do – first on a global level, then on a local level, and finally on a personal level.

Climate change is the issue of our time. It will define how the twenty-first century plays out, and it will shape our collective future for centuries to come. Right now, we are at a perilous point in human history. No generation has ever faced such a threat. Many times in the past, when regional civilisations have risen and fallen, environmental damage has had a significant role to play. This time, it is no regional civilisation at risk – it is the whole of our global community. If we cannot get on top of the problem of climate change in the next decade, we are facing a grim future.

The good news is that we can get on top of it – if we act quickly.

We know exactly what to do to avoid the bad stuff and get to a future that is safer, cleaner and better for everyone.

I

THE GLOBAL

1

Earth's ever-changing climate

When I cast my mind back to the summers of my childhood in Canterbury, they're filled with long, hot days drenched in sun, my bare feet running over scorched grass. There are trips to the river, the pool and occasionally the beach – anywhere to cool off.

Now, I live on the Kāpiti Coast and I work in Wellington, which isn't exactly famed for its favourable summer weather – but, even making an allowance for this, I could swear there are quite a few more wet and cold days than there were back when I was a kid. It doesn't always feel to me, when I look out the window, like the climate is warming. So can it really be true – are the days actually getting warmer?

Well, to answer that question, I first have to point out the blindingly obvious: memory is unreliable (and not just mine, to be fair). My memories might tell me how my life has changed, but they can't really tell me anything reliable about how the weather has changed. When I was a child and it was sunny, I spent every moment I could outdoors. Now, rain or shine, summer or winter, I spend a lot more of my time inside. I come into contact with the weather only in short bursts, either when I must go out in it or when I choose to. So it's only natural that it feels like there was more sunshine and warmth when I was a kid – in a sense, there was, but it's only because I was outside enjoying it more often.

Trying to remember what the weather was like many years ago is difficult. It's one thing to recall that it was raining last Wednesday, entirely another to pinpoint what was happening on the first Wednesday of October three years ago. It's similarly difficult to track the way the weather changes from one week to the next, let alone the climate from year to year or decade to decade. Sure, we can look outside and see the clouds some days or sunshine others, but that just tells us about the weather. It doesn't tell us much – if anything – about the climate and how it's changing.

And that brings us to the answer itself: yes, it's true. The climate is indeed changing, and the days actually are getting warmer. For more than a century, average daily temperatures have been rising across the globe. The days are now hotter on average than they were 30, 40, 50 years ago and more. There's no dispute about it; it's a fact.

This change is happening right over our heads, but it's all on such a grand scale that we need to do more than look out the window to see it. We need to be able to zoom out and up and away and get a look at the world as a whole. We need to be able to think about time on Earth's scale – one that covers decades and centuries and millennia and all the years in between – not within the blink-and-you'll-miss-it confines of our own unreliable memories or the narrow picture held within our home window frames.

*

Back in the 1970s, when I was at secondary school, we were taught that the climate was simply 'the sum of the weather'. You took all of the weather experienced in a particular area over a long enough time – a few decades, say – and averaged it out to define the region's climate. I learnt a little about the weather, and I learnt that some places were warmer and some cooler, but that was about it. There was very little 'why', and no discussion of how climates can and do change.

Researchers at the time knew the climate varied, and had been studying these variations since the nineteenth century, but these changes were cast mostly as historic events, disconnected from the weather and climate of the present day. At school, I didn't need to know about any of that. As long as I could cite the annual mean temperature in Kuala Lumpur, I was good for my exams.

Thankfully, in the decades since, both I and the scientific community as a whole have learnt so much more about the climate. These days, it is seen as a dynamic, constantly changing system, where the past is in conversation with the present and gives us hints to the future. The weather is part of this system, but there are other bigger, slower forces at play, too – internal variations like El Niño, and changes that result from variations in sunlight levels and greenhouse gases. These things give an individual character to our seasons and years, but they mostly operate independent of the day-to-day weather.

So, while the climate is governed by the same physical laws as the weather, it operates on a much broader scale and over longer time periods. Rather than being the sum of the weather, the climate is more like a slowly changing background. It sets the scene for the weather we experience. And, over the years and centuries, variations in the climate will shape and reshape the weather.

The climate system encompasses everything. There's the flow of the winds and the weather, the global ocean circulation, the rivers, the lakes, the land surface, the glaciers and ice sheets and other frozen parts of the world. There are the changes from day to day, from month to month, from century to century and further back. There is all life on Earth – life which both supports and is supported by the climate. What grows where, especially in the tropics, has an effect on the composition of the atmosphere and how it flows around the globe, which in turn affects where and how much it rains, and how much temperatures vary.

There are many beasts that come and go in the climate menagerie, and each of them pushes and pulls on our weather systems for a month, a season or a year at a time. Warm-blooded El Niño and its cooler counterpart, La Niña, are climate patterns anchored in the tropical Pacific Ocean, but they can affect the weather worldwide. The Southern Annular Mode (SAM) encircles Antarctica, controlling the winds and the tracks of storms over the Southern Oceans. In the climate system everything is connected to everything else, and that's especially true in the tropics, where so much energy is held in the climate system; what happens in the tropics definitely does not stay in the tropics.

Then there's us. Humans. Right in the thick of it. Just like every other form of life on Earth, we too are part of the climate system, affected by and affecting it. What we do to the climate, we do to ourselves. The climate, the environment, is not something separate from us, not something we can manipulate or trade on. We are in it and part of it. If the climate changes, we change.

Climate change is about much, much more than things just getting warmer.

*

One way of looking at the climate is to see it as a constant interplay of many moving parts: fast and slow, cold and hot, forces that work together and others that oppose each other.

The fast atmosphere is always buzzing, with storms and sunshine coming and going, and winds moving air around the whole globe in a matter of days. The atmosphere moves heat around the world, but it does so speedily. A bit like us, it has very little 'memory' from one season or one year to the next.

The oceans – our largest and least-known ecosystem – also carry heat around the world, and away from the tropics, but they do so ponderously, especially in the deeper areas. Snow and ice, and how much of it we have, also affects our weather. It's in these slow-moving but powerful bodies that we find the climate's 'memory': the changes in our climate over seasons, years and centuries are dictated by changes in the oceans' circulation and in snow and ice cover on land.

These fast and slow elements are always talking. The fast atmosphere brings us our daily weather – what we see when we look out the window – but its behaviour is informed by the oceans and land below.

As well as the fast atmosphere affecting the slow oceans (and vice versa), there's another important relationship thrown into the climate mix: the difference in temperature between the warm tropics and the cold poles. This difference drives both the weather and the climate, and is a bit like the volume knob on an amplifier: wind up the difference, and winds pick up speed, storms become stronger; reduce the difference, and winds drop away, storms become milder.

Those winds aren't just a side-effect. They have a very important job of their own: to try to make the temperature the same everywhere, all over the globe. They're relentlessly busy, blowing warm air towards the poles and bringing cold air towards the Equator, but their job isn't straightforward. The rotation of the Earth complicates their passage, meaning they end up rotating around high- and low-pressure centres, instead of just flowing from one place to another.

All the weather we experience – every storm or rain shower, every sunny day – is part of an eternal quest to equalise temperatures between the tropics and the poles. Beneath the ocean, the same thing's happening with the currents. All that effort is ultimately futile, however, because with every new day the sun douses the tropics with warmth, while the frigid poles get a pittance. As long as the sun shines, the dance goes on.

*

Sometimes, when I give public talks on climate change, I'll open by promising my audience a guaranteed 100 per cent accurate forecast for the rest of the year.

'The winter months,' I'll say, 'will be colder than the summer months.'

The laughter is sometimes a bit pained, and occasionally an audience member will point out the obvious. 'That's cheating!' they'll say. 'Everyone knows the seasons change. That isn't a forecast!'

Many of us take the seasons for granted. We don't stop to wonder why they occur … or whether they will. Yet the forces that cause the seasons to change are, in many ways, the same forces that cause longer-term climate change. Both are consequences of how much energy falls on the Earth. Neither is about the weather.

In fact, the seasons are on their own a kind of recurring 'climate change'. The daily weather varies, but as we move into spring and summer the days get warmer, then begin to cool again through autumn and into winter. The change is a result of how much sunlight is falling on our part of the world. In the summer, the sun is more directly overhead so the sunlight is more intense, the days are longer, the temperatures higher. In winter, the sun sits lower on the horizon and sunlight is spread more thinly on the land, so the days are shorter and temperatures lower. These changes through the year are very regular and predictable – and, if that was all that was going on, each year would be much like every other.

While the seasons change on repeat from year to year, they don't in themselves change the overall climate. When it's winter here, it's summer in the Northern Hemisphere, and vice versa. When you average that out across the globe and through the year, you get the same global temperatures each year. It's much the same with other climate patterns. El Niño makes global temperatures a little higher, but it does so at the expense of making the global oceans a little cooler. Meanwhile, La Niña cools the atmosphere and warms the oceans slightly. When the amount of heat energy in the atmosphere *and* the oceans is tallied, there is in fact no change through the El Niño–La Niña cycle. The same goes for all the 'natural' variations in the climate. Heat can move around, making one region warmer at the expense of cold elsewhere, but no heat is added or subtracted in total.

Sometimes people say to me, 'You can't forecast the weather two weeks ahead, so how can you possibly forecast the climate a hundred years ahead?' and, when they do, I think of the seasons. Forecasting the climate is more like forecasting the seasons than it is the weather. I don't need to know what the weather will be in Wellington each day in July in order to be able to tell you that July will be cooler than January. And, in just the same way, I don't need to know the precise weather on New Year's Day 2100 to be able to tell you that the whole globe will be warmer in 2100, if the climate keeps changing as it is today.

It would be great to know what the weather is going to be in January of 2100, but that is quite simply impossible. American meteorologist Edward Lorenz was running simplified models of the climate back in the 1960s when he stumbled on the chaos in the weather. His work – including the discovery that small changes in present conditions produce big changes days and weeks into the future – led to the understanding that the sequence of weather is predictable for a week, maybe two, but after that the best we can do is guess (despite what some might like to claim). This element of randomness in both the weather and the climate means weather and climate patterns don't always behave the way that they 'should'. Sometimes, they just do their own thing. A season or a year can be wetter or cooler than average for no identifiable reason – it's just the way things unfold.

Other times, though, there's an extremely obvious reason.

*

In April 1815, Mount Tambora in Indonesia erupted, blowing away the top 1,500 metres of the mountain, killing tens of thousands of people and spewing so much ash into the atmosphere it stole summer from the entire globe the following year. It was the most powerful eruption for at least 1,000 years, and one of the most powerful of the past 10,000 years.

The 'year without a summer' that ensued was a result of the volcano's sulphate cloud getting in the way of the sunlight, resulting in unseasonably chilly and unsettled weather across the Northern Hemisphere, major crop failures and the century's worst famines. The general gloom was unavoidable, and dystopian visions made their way into art, leading to works including Mary Shelley's *Frankenstein*.

Over 60 years later, Krakatoa – also in Indonesia – erupted and prompted another period of global cooling that saw record snowfalls across the planet. The vivid sunsets brought about by Krakatoa's sulphate clouds also made their way into the art of the 1880s, including a number of English watercolours and possibly the disturbingly lurid sky in Edvard Munch's 'The Scream'. Notably, this eruption was one of the first major geophysical events to occur after the invention of the telegraph, meaning a lot of observational data was gathered in near real-time, spurring on early research into the structure of the atmosphere.

The sun provides all the energy that drives the weather and the climate. If the sun's brightness changes, so does the climate. Volcanic eruptions like that of Mount Tambora are an example of this – when they're significant enough, they fire sulphate particles into the stratosphere, which linger there for a year or two and reduce the amount of sunlight making its way to Earth. Major eruptions essentially make the sun dimmer at the Earth's surface, and the whole globe cools as a consequence. And, as we've seen through art, that dimming doesn't just affect the climate, but humanity's general mood and state of being as well.

This cooling effect is especially marked when the volcano that erupts is located in the tropics. Then, the sulphate cloud can spread out into the northern and southern hemispheres. After both Mount Agung in Indonesia erupted in 1963 and Mount Pinatubo in the Philippines in 1991, global temperatures decreased by half a degree or more for a couple of years. The eruption of Mount Pinatubo, in particular, in 1992 gave New Zealand its coldest year since the 1940s.

A curious fact about sunlight is that, although it affects the surface temperature of the Earth, it hardly warms the air at all. Some sunlight is absorbed into the ozone layer and some is reflected away by clouds, but the rest just passes right through the air and hits the Earth. This means that the Earth warms and, just like the sun, radiates energy into space – but Earth obviously radiates a much less intense form of heat, what we call infrared energy.

*

When it comes to the Earth's surface temperature, the sun is just one of two important players. The other? Greenhouse gases.

Around two centuries ago, scientists discovered these gases, which absorb the infrared energy the Earth radiates – the main ones being carbon dioxide, water vapour and methane. They act a lot like a blanket on a bed. If you put a heavier blanket on your bed, you'll be warmer during the night; similarly, put a 'heavier' blanket of greenhouse gases on the Earth, and its surface will be kept warmer, because a greater amount of infrared energy is radiated back at it.

The warmest parts of the atmosphere are those closest to the ground. This is because the air is warmed from below, and it's also because greenhouse gases (especially water vapour) are most abundant near ground level. When you heat a pot of water on a stove, you see bubbles rising from the bottom, where the heat source is. It's similar in the lowest 12 kilometres or so of our atmosphere: when water or air is heated from below, what's at the bottom becomes more buoyant than what is above, and there is a natural tendency for rising motion. In the atmosphere, instead of bubbles, we see updraughts and thermals, which ultimately form clouds and rain.

All that said, greenhouse gases comprise less than 1 per cent of the atmosphere (not counting water vapour). The remaining 99 per cent is made up of nitrogen and oxygen, which are almost completely transparent to sunlight and infrared energy, so play virtually no part in heating the Earth.

However, don't be misled – greenhouse gases might only make up a tiny proportion of our atmosphere, but they're by no means insignificant. If our blanket of greenhouse gases wasn't there, temperatures at ground level would be well below freezing, and we wouldn't be here at all. In fact, if the air wasn't heated from below and if greenhouse gases didn't work the way they do, the weather would hardly be here either. There would be very little rising motion in the air, so there'd be very little in the way of clouds or rain.

*

The Earth has not always been the way it is today. Fifty million years ago, there was virtually no ice anywhere, sea levels were around 70 metres higher than they are now, and crocodiles and turtles swam at the North Pole.

By contrast, around 20,000 years ago, sea levels were about 120 metres lower than today and there was so much water locked up in ice worldwide that the three main islands of Aotearoa were one big island. You could walk from Kaitaia to Rakiura/Stewart Island without getting your feet wet. In fact, for most of the last couple of million years, that is how New Zealand has been. The present configuration is the anomaly.

As far back as we can look in the geological record, the Earth's climate has been changing in response to levels of sunlight and greenhouse gases (particularly carbon dioxide).

Over the course of tens of thousands of years, the amount of sunlight that falls on the Earth has risen and fallen as a result of changes in both the shape of the Earth's orbit and our angle of tilt. Especially on the big Northern Hemisphere continents, the sunlight has essentially become gradually brighter, then gradually dimmer, then gradually brighter again. When it has dimmed, the climate has cooled, massive ice sheets have formed, and the oceans have inhaled carbon dioxide. When the sunlight has brightened, the climate has warmed, the ice has melted and the oceans have exhaled carbon dioxide. It's this process that causes the ice ages our planet has seen come and go so many times before. As the Earth has moved through this cycle, average temperatures have gone up and down by as much as 6°C, and sea levels have risen and fallen by 100 metres or more.

Our current series of ice ages began about two and a half million years ago, and it took around 10,000 years for the Earth to transition from the depths of the last ice age to the current interglacial period, known as the Holocene. The transition wasn't smooth – it came in fits and starts, with ice dams suddenly collapsing, flooding water into the oceans and raising sea levels by metres in only a few decades – but the climate and the sea levels eventually stabilised around 10,000 years ago. Since then, they've remained nearly constant.

It's been the longest period of climate stability for many millions of years.

*

Scientists have been developing an understanding of past climates for about as long as they've been studying our present one. The first inkling of past ice ages came from 'erratics', large boulders that seem completely out of place in relation to the ground they're sitting on. In the nineteenth century, it was suggested that receding glaciers might have moved these huge chunks of rock to their present-day locations – and, if that was the case, then there had to have been an ice age at some time in the past. Trying to work out what could have caused such an ice age was what led scientists to understanding the impact of greenhouse gases and sunlight on our climate.

Further understanding of the climate and all its ice ages has come from the Earth itself. When snow crystals settle on the surface of an ice sheet or glacier, small pockets of air form between them. As the snow collects and becomes a body of ice, these pockets of air remain trapped in the ice. By drilling down into the ice and extracting frozen cores, scientists can analyse those ancient bubbles and learn the story of the time when they first formed.

That might all sound fairly straightforward, but it's not. Not only do the scientists who extract these cores have to set up camp in Antarctica or Greenland, they also have to spend months in these hostile environments, drilling holes in the ice, carefully removing the frozen cores and safely transporting them back to the laboratory. Then, the detective work begins: long hours spent painstakingly analysing what the core contains in order to decipher its story.

Here in New Zealand, the science happens at the National Ice Core Research Facility at GNS Science in Lower Hutt. There, scientists melt the ice from the bottom up and capture the gases contained in the air bubbles as they are released. Measuring how much carbon dioxide or other greenhouse gases are in a bubble of air is fairly simple; the trickier bit is working out the temperature at the time the air was trapped, and the age of the air bubble itself.

The temperature part is quite ingenious. Different kinds of oxygen isotopes have different weights, and the precise combination of these different flavours of oxygen is a consequence of temperature. So, if you measure the different oxygen isotopes and their relative abundance in the air bubbles, you can work out the temperature of the air they must have been in.

The age part comes down to a number of factors. In the upper part of a core, annual layers are visible – this means we can count back through the years much like we do with the rings in a tree's trunk. A big volcanic eruption is a helpful time marker, as it deposits a layer of ash at a well-defined moment in the past. Eruptions from New Zealand volcanoes such as Mount Tarawera and Lake Taupō, for instance, have left discernible layers of ash in Antarctic ice. Going farther back in time, radioactive decay becomes useful. Dust that has blown onto Antarctica in the past contains isotopes of radioactive substances such as uranium, which can be used to work out the age of the dust and hence the ice it's frozen into.

The longest ice cores, drilled into the thickest parts of the East Antarctic ice sheet (over 3 kilometres deep), can take us back in time around 800,000 years, through the last eight ice age cycles. That's where the ice record stops, for now. There are plans to drill longer cores in Antarctica by going on a bit of an angle instead of straight down, and that may take us back over a million years – but that's about the limit. To work out what was happening more than a million years ago, we must turn to the geological record, and that means drilling through ocean-floor sediments.

The sediment record contains the shells of tiny sea creatures (called forams), fossils of plants and even smears of DNA, all of which can transport us much farther back in time than bubbles of air can – up to 100 million years or so. And, while we can't discern year-to-year climate variations that far back, we can gather a picture of changes that have happened over centuries or thousands of years. The isotopes of oxygen and carbon in forams' shells, for instance, tell us about the water temperature when the creatures were alive, as does the presence or absence of different species. It's in sea-floor sediment that we can track changes in global temperature and atmospheric greenhouse gases before the extinction of the dinosaurs, as well as the slow descent into the current era of ice ages of the last two-and-a-half million years or so.

For years, ocean-drilling has been coordinated internationally by the International Ocean Discovery Program. And, if drilling ice cores sounds challenging, drilling through the sea floor is harder again. First of all, your coring gear has to be able to reach all the way down to the sea floor before it can even begin drilling, and you'll need a very special kind of ship – something like the *JOIDES Resolution*, which is basically a floating town with a massive drilling rig in the centre. In deep water, it can take half a day just to get the drill bit down to the sea floor. Then, once the drilling starts, special thrusters keep the ship in place over the drill hole and stop it from rotating while the drill does.

The Antarctic Drilling Project, known as ANDRILL, took things up a notch by drilling through both the ice *and* the sea floor at once. A drill rig was set up on top of the Ross Ice Shelf, a massive block of floating ice about twice the size of New Zealand and several hundred metres thick. The scientific team then drilled almost a kilometre down through the ice – which they weren't interested in at all – to the sea floor beneath, and extracted over a kilometre of sediment core containing records of the past 20 million years or so. This core showed that the Ross Ice Shelf has come and gone many times in the past, melting away in warm periods and re-emerging in colder times. This was a big shift in our view of Antarctica: where once the icy continent had been seen as fairly static, it was shown to in fact be a sensitive and dynamic environment. The West Antarctic, in particular, is very sensitive to warming, and especially to the sort of ocean warming happening today.

The ability to extract and read ice and sediment cores is one of the major scientific achievements of the past century.

Doing so grants us an understanding of Earth's climate across tens of millions of years, an unfathomable time span during which mountain ranges have risen and worn away, whole continents have shifted, and massive volcanic eruptions have poured lava over vast areas. It's thanks to the first ice cores that were drilled in Greenland, back in the early 1990s, that we know the shape of the last few ice ages, as well as about the stability of the last 10,000 years.

The book of the climate is written into the surface of the Earth itself, and right now humanity is in the process of adding a new chapter. There are two things that influence our planet's surface temperature – the sun and greenhouse gases – so one of them has to be the culprit. I can tell you it's not the sun, because the sun's not getting any brighter. In fact, over the last 50 years, it has actually become a bit dimmer. Meanwhile, greenhouse gas levels are increasing rapidly. Why? Because we've been busy pumping them into the air with increasing abandon.

As a result, Earth's ever-evolving climate is undergoing its latest change, but this one is unlike any before it. This one is human-driven.

2

A complicated relationship

When I was a kid, steam engines ruled. I grew up in a railway family in a railway town, and the local rail yards weren't far from our house. There, I would watch locos coming and going, being checked and repaired, and coal wagons would arrive from the West Coast heading for Christchurch, changing places with loads of grain and produce from the Canterbury farmers.

When the wheels started to turn in a steam locomotive, massive rods and pistons would move back and forward, up and down, and steam would come shooting out like dragon's breath. The wheels would often skid before they gripped the tracks and started moving the engine forward. Sometimes, I got to go with Dad into the main room of the station, where he oversaw the operation of the yards, constantly switching the points to map rolling stock from one set of tracks to another. Once or twice, I even got to help change the points myself, grabbing the metal levers in my then-tiny hand and pushing them forwards or pulling them backwards. It was all so exciting. These powerful engines were so big, so alive, but they moved along the tracks in such a silky, almost delicate fashion, in spite of the tons of coal and water they hauled.

By the 1970s, the era of coal-powered steam locomotives had come to a close in New Zealand, but not the use of fire to create power. I thought about this when I woke up this morning and my home was a bit chilly. As I lit the fire and watched the flames lick at the logs, it struck me that I was doing something humans have been doing for pretty much our entire time here on the Earth. There is, in some ways, nothing more human than the desire to light a fire and watch it burn. Fire keeps us warm, protects us, provides us with light and comfort. Today, combustion also powers almost all of our transport, from a car trip to the shops to a flight across the world. It even provides a good fraction of the world's electricity. And, for all these things, it's energy-dense fossil fuels like coal or oil that we burn – they've provided us with the most valuable forms of fire we have ever known, and have transformed our societies and economies.

The trouble is, of course, that burning all these fossil fuels has come at a terrible cost: in mere decades, we've managed to take millions of years' worth of stored carbon and pump it back into the atmosphere. So, we might have harnessed fire power and made amazing progress with that technology, but we've also simultaneously set about the Earth's next climate change.

I'm not the only one to remark on our complicated relationship with setting things alight. In an essay for *The New Yorker* in March 2022, American author and climate activist Bill McKibben succinctly summed things up as follows:

Our species depends on combustion; it made us human, and then it made us modern. But, having spent millennia learning to harness fire, and three centuries using it to fashion the world we know, we must spend the next years systematically eradicating it. Because, taken together, those blazes – the fires beneath the hoods of 1.4 billion vehicles and in the homes of billions more people, in giant power plants, and in the boilers of factories and the engines of airplanes [and] ships – are more destructive than the most powerful volcanoes, dwarfing Krakatoa and Tambora. The smoke and smog from those engines and appliances directly kill nine million people a year, more deaths than those caused by war and terrorism, not to mention malaria and tuberculosis, together. (In 2020, fossil-fuel pollution killed three times as many people as COVID-19 did.) Those flames, of course, also spew invisible and odorless carbon dioxide at an unprecedented rate; that CO_2 is already rearranging the planet's climate, threatening not only those of us who live on it now but all those who will come after us.

That essay was titled 'In a World on Fire, Stop Burning Things'. Indeed.

*

Back in 1958, American scientist Charles Keeling started collecting carbon dioxide samples at Mauna Loa Observatory, an atmospheric baseline station situated on the volcano of the same name in Hawaii. Within a few years he had collected enough data to show that levels were rising steadily – and they've continued to do so since, in what is now known as the Keeling Curve. The rise during that time has accelerated, too.

When the Keeling Curve started, the increase was around one part per million per year, but now it's up to between two and three parts per million per year this century.

To put this in further context, when the first coal-fired steam engines went into service back in the mid-1700s, annual carbon dioxide emissions were around 10 million tons per year. By 1800, they were up to 30 million tons. Fifty years later, they'd jumped to 200 million tons. These days, they are closer to 40 billion tons per year. For well over a century, global emissions have been rising nearly exponentially.

We know that this increase is a consequence of burning because, at the same time as carbon dioxide levels are rising, oxygen levels are going down. (No need to panic about suffocating, though, as the amount of oxygen in the air is hundreds of times the amount of carbon dioxide. We'll have a lot of other problems to deal with before running out of oxygen is ever an issue.)

Furthermore, we also know the carbon being added to the atmosphere is fossil carbon because the fraction of radioactive carbon in the atmosphere is decreasing. Fossil fuels really are fossils – the remains of plants from hundreds of millions of years ago that have been transformed by heat and pressure in the Earth over millions more years – and they have been underground so long they no longer contain any radioactive carbon atoms. So, when they are burned, it's a depleted form of carbon dioxide that's released, and there's a corresponding decrease in the fraction of radioactive carbon in the atmosphere. This is exactly what's happening in our atmosphere, at almost exactly the rate we would expect based on how much coal, oil and natural gas is burned every year.

*

So, why all the angst over carbon dioxide?

All greenhouse gases matter, but scientists pay most attention to carbon dioxide because it's the one we emit the most of, and it's one of the ones that sticks around the longest. Once it's emitted, it'll remain in the atmosphere for centuries. So, even if you decrease your carbon dioxide emissions tomorrow, you're still going to suffer the consequences of the stuff you already emitted for hundreds of years to come.

One way to think of our blanket of atmosphere, with its sprinkling of greenhouse gases, is as a very thin skin that sits over the Earth. If the Earth was the size of an apple, the atmosphere would be about as thick as the apple's skin. There is nothing holding the atmosphere in (a thought that, I confess, worries me a little); gravity is all that stops it from flying away completely. Most of the mass of the atmosphere is in the bottom 10 kilometres or so, and at around 100 kilometres altitude the atmosphere is so thin it is basically outer space. There is really very little between us and space – just that thin skin of air, which determines the whole of the climate and the fate of all living things on Earth. What might seem like small changes in this thin layer of air can have very big consequences for all of us.

Carbon dioxide might make up only a tiny fraction of our thin layer of atmosphere, but it is absolutely critical to energy flows on Earth. Without it, we would not be here. And, as study of past climates has shown, if the balance is disturbed there will be consequences. The arrival of *Homo sapiens* and the development of our fossil-fuelled civilisation has been Earth's biggest tipping point since an asteroid crashed into it 66 million years ago. Prior to the Industrial Revolution during the eighteenth century, carbon dioxide emissions were very low, and grew only slowly at first. But the growth has been exponential recently, up to nearly 40 billion tons per year. That means that, in just a couple of centuries, we've increased carbon dioxide levels by nearly half. This is the biggest change in over three million years, and the fastest for tens of millions of years.

Here, it's important to note that that not every human – and not even every country – emits equally. What those total emissions numbers hide is the fact that some of us are emitting significantly more carbon dioxide than others. Differences in standards of living and levels of consumption vary wildly from place to place, and so too do the carbon emissions each country is responsible for.

The populations of most developed nations consume way more resources and emit many times the amount of carbon dioxide than people in the poorest countries of the world do. To put that in perspective, the average person in New Zealand or the US might be responsible for somewhere in the vicinity of 15 tons of carbon dioxide emissions per year – that's three times the global average. Meanwhile, a person from one of the poorest countries in the world might account for only 100 kilograms of emissions per year – less than 1 per cent of what the average New Zealander is responsible for. In other words, in three days, the average Kiwi can emit more greenhouse gas than the average person in Mali does in a whole year.

*

You may have heard that, actually, there are other greenhouse gases that are more important than carbon dioxide. Water vapour, for instance.

It's true that water vapour is a big issue – but it's not *our* issue, because we don't directly control how much of it is in the air. The temperature does. When the climate warms, more water evaporates at the Earth's surface, leading to more water vapour (or moisture) in the air. There is an exponential relationship between air temperature and the maximum amount of water vapour in the air – this means the amount of moisture in the air goes up very rapidly with temperature. So, the more the climate warms, the more water vapour we end up with in the air.

The thing about water vapour, though, is that it doesn't stay in the air for very long. If it is added, the air pretty quickly becomes so saturated that moisture is forced to condense back to liquid water, forming clouds and rain. This means water vapour on its own cannot drive warming, as it falls out before it can affect global temperatures. Even if every person in every country all over the world boiled their jugs at exactly the same moment, the water vapour added to the air would saturate the air in most places, leading to more clouds and maybe more rain, but not higher temperatures. Personally, I find it immensely comforting to know that, in the grand scheme of things, making a cup of tea isn't going to add to global warming.

While we don't have a direct impact on water vapour itself, we are directly affecting its main driver – the temperature –

by pumping carbon dioxide and other greenhouse gases into the air. As the climate warms, water-vapour levels increase, and water vapour is indeed a powerful greenhouse gas. In an effect known as 'water vapour feedback', it effectively doubles the warming we would get from increasing carbon dioxide alone. So, while it can't warm the planet on its own, water vapour does amplify human warming of the planet.

Another greenhouse gas you may have heard about is methane – you know, the stuff that mostly comes from cows and sheep burping. Methane is a hot topic here in New Zealand, due to our agriculture industry. As of October 2022, and using the standard way of comparing methane and carbon dioxide, Statistics NZ reported that methane made up 43.5 per cent of the country's gross greenhouse gas emissions. By comparison, carbon dioxide comprised 43.7 per cent, and nitrous oxide was 10.7 per cent. Of those methane emissions, nearly 90 per cent were produced by livestock, and most of the nitrous oxide came from agricultural soils, specifically the urine and dung deposited by grazing animals.

Those are sobering numbers, and there's no doubt that agricultural emissions are a real concern for a country like New Zealand. However, comparing carbon dioxide and methane emissions requires a bit more information than the gross numbers. Methane stays in the air, on average, for only around 12 years. This means that, if methane emissions decrease, the amount that's in the air will also start decreasing pretty promptly – very different from carbon dioxide, which will still be there having an effect long after we stop emitting it. While it's up there, methane is also much better at absorbing heat than carbon dioxide is.

So how then do we compare the relative effects of these gases on climate change? For the last few decades, emissions have usually been measured in carbon dioxide equivalents (CO_2e). Basically, you pick a specific time period – the next 20 years, say, or the next 100 years – and work out how much warming you'd get from a kilo of carbon dioxide in that time. Then, you measure other gases like methane or nitrous oxide in comparison to that. The standard timeframe used in national reporting and international negotiations is 100 years, and over that time period a kilo of methane is around 25 times as powerful as a kilo of carbon dioxide – but the thing is, the methane will disappear in that time, while the carbon dioxide will linger for hundreds more years. If you choose a period longer than 100 years, methane suddenly appears less important. So, comparing the gases in this way can be a bit of a fiction.

Ultimately, all the greenhouse gases are different. Each one has a specific lifetime in the atmosphere, and specific abilities to absorb heat. It's this complexity of comparing them, and their relative impacts, that has led in recent years to split-gas targets for emissions reductions – targets that treat different gases differently. For instance, New Zealand's Climate Change Response (Zero Carbon) Amendment Act 2019 stipulates reducing our net carbon dioxide emissions to zero by the year 2050, but reducing biogenic* methane emissions to only 24–47 per cent below 2017 levels in that same timeframe.

While it's obviously a good idea to focus on reducing all greenhouse gas emissions over time, it's carbon dioxide that matters the most right now. That's the gas that's going to be with us the longest, and that will therefore have ongoing impacts on our climate – both here in New Zealand, and globally. First, we need to get our carbon dioxide emissions down. Then, we can really focus on methane and the other gases, too.

Another claim that sometimes gets made is that one decent volcanic eruption would add more carbon dioxide to the air than all the emissions humans have ever produced. That might sound impressive, but it's completely wrong. While volcanoes do emit carbon dioxide, the emissions of all the volcanoes in the world add up to less than 1 per cent

---

* Produced by biological (plant and animal) sources.

of the annual emissions from humans burning fossil fuels. We would have to go back to the late 1700s, when fossil-fuel burning was in its infancy, to find a time when volcanoes were doing more emitting than we are right now. When Mount Pinatubo erupted in 1991, for instance, there was no sign of a spike in carbon dioxide. Same story with Mount Agung in 1963. The global atmosphere didn't notice at all, in terms of carbon dioxide.

*

The world has known about climate change and the role of human emissions of greenhouse gases for a long time, and tackling climate change has been on the agenda of world leaders for more than three decades.

To put this in perspective, here's a brief timeline of key events:

- In 1988, the Intergovernmental Panel on Climate Change (IPCC) was created to assess the science related to climate change and provide governments with reliable information to develop climate policies.
- A few years later, in 1992, the United Nations Framework Convention on Climate Change

(UNFCCC) established a treaty signed by 154 states promising to combat 'dangerous human interference with the climate system', in part by stabilising greenhouse gas concentrations in the atmosphere.

- In 1995, the UNFCCC held its first COP (Conference of the Parties) meeting in Berlin, Germany, to review the implementation of the treaty, and since then the COP has met every year in different countries.
- The UNFCCC negotiations eventually led to first the 2007 Kyoto Protocol, then the 2015 Paris Agreement. The central aim of the Paris Agreement is to strengthen the global response to climate change by keeping global temperature rise this century well below 2°C above pre-industrial levels, and ideally to limit it to 1.5°C.

So, since the 1980s, a lot of conventions and treaties and agreements have occurred … and in that same time, global emissions have nearly doubled. We have made the problem twice as bad. Half of all the carbon dioxide that's gone into the atmosphere since the 1700s has been emitted since about 1990. Put another way, half of all global carbon dioxide emissions have been made during Taylor Swift's lifetime – and half those since she released her first album.

I receive many emails from people wanting to discuss – or, unfortunately, argue with or dismiss completely – the science of modern-day climate change. Quite a number claim to have a background in geology. 'The climate has always changed,' they tell me, citing the ice ages and the many enormous changes our Earth has seen over millions of years.

Geologists are the historians of the climate system. They study the Earth, and that means they must also understand how the climate has changed in the past. But geology is all about million-year processes, during which mountain chains grow and weather away, the Earth's orbit changes, and the sun grows brighter and dimmer and brighter again. To a geologist, 10,000 years is the blink of an eye. Massive forces are at work over millions of years, changing the shape of the climate and of the very Earth itself.

Humans, on such a majestic stage, are both puny and latecomers. In order to try to get his head around our existence in terms of the age of the Earth – 4.5 billion years – author John Green uses the analogy of a calendar year, with the formation of Earth being 1 January and today 31 December at 11.59pm. 'The first life on Earth emerges around February 25,' he writes in *The Anthropocene Reviewed*. 'The meteor impact that heralds the end of the dinosaurs happens around December 26. *Homo sapiens* aren't part of the story until December 31 at 11.48pm.'

I've been fortunate to work with many of New Zealand's finest geologists and, in order to communicate effectively, we've each have had to do a bit of translating and adjusting of perspectives. It turns out knowing about the weather doesn't help you understand the ice ages! But, as important as Earth's geology is, I believe too deep a focus on humanity's supposed 'insignificance in the scheme of things' can sometimes get in the way of realising just how potent our activity can be. Yes, the usual things that change the climate are vast and inexorable, and yes, we are latecomers – but, over little more than a century, humans have managed to do the work that would have taken tens of thousands of volcanoes a million years or more. In a matter of decades, we have undone millions of years of geology.

One of the difficulties in attempting to fully comprehend human-driven climate change arises from the need to bear a series of wildly different timescales in mind. To really grasp the magnitude of what's going on, we need to understand just how quickly we have emitted carbon dioxide into the atmosphere. But, to understand the far-reaching consequences, we must simultaneously think slower, take a wider view. It will take many centuries for the nearly two trillion tons of carbon dioxide we have dumped into the atmosphere in only 50 or 60 years to be gradually absorbed into the oceans and through geological processes.

The issue, of course, is that a human lifetime is much shorter than the time periods that matter for the Earth. Melting an ice sheet takes place over thousands of years. Getting heat to fully penetrate from the ocean surface all the way to the sea floor takes centuries. We're not going to see those things happening when we look out the window – but they are happening.

3

The climate now

'Why is such a small change such a big deal?' I once had a student ask. 'I wouldn't notice if the water in my shower warmed by a degree.'

He wasn't wrong. I doubt if anyone could detect a 1°C change in the temperature of their shower water. But, if that change came in the form of short bursts of scalding water with normal water in between, I know I would definitely notice. Especially if those scalding bursts started happening more often.

That is effectively what is happening around the world. The weather is 'normal' much of the time, but occasionally it's unusually warm or wet or dry – and those unusual occurrences, or scalding bursts, are becoming more usual over time. We don't feel climate change as a gradual warming over decades, but in the change in the occurrence of extreme events.

Since the middle of the nineteenth century, we've been able to track the trend in global air temperatures, outside of the background ups and downs of usual climate variability, and on average the days now are a little more than 1°C warmer than they were at the end of the nineteenth century. However, the rate of warming is not consistent from place to place. Over the oceans around Antarctica, for instance, there has been almost no warming in the last 70 years. Meanwhile, the Arctic and far-northern countries have warmed by 2°C or more. New Zealand sits somewhere in between these extremes – our country has been warming at about the global rate.

That air-temperature change also represents only a tiny fraction of the total warming that has happened to the climate system over the last century. A little of the heating has gone into melting ice and a little has gone into warming the surface of the land, but most of it has gone into warming the oceans. Water covers nearly three-quarters of the Earth's surface, and it is very good at absorbing energy – much better than the land or air. It's estimated that over 90 per cent of the total heating that's accrued on Earth since we started adding greenhouse gases to the atmosphere has gone into the oceans.

And the temperature isn't the only thing we're seeing change, either. Ice is melting, sea levels are rising, and the warming oceans are also becoming acidic. The land is heating up more in some places than others. We're hit by increasingly severe weather events in the form of floods, fires, heatwaves and droughts. And all of this is having a knock-on effect for the flora and fauna that call the world's various environments home – including us.

*

Through much of recorded history, rivers and oceans have been our highways. We have set out across the wide seas in search of other places and new experiences. We have used rivers as a means of transport and industry. We have built settlements and cities near coasts, in natural harbours and around river mouths. And the connection we feel with the world's waterways is more than transactional. This is especially true of our oceans. For many of us, standing on a beach while the salty wind whips our face and the waves crash at our feet is both a grounding and a spiritual experience. It's a reminder of the natural world, and our place in it. The water stretches to the distant horizon, meeting with the sky and disappearing further than our eyes can follow, with such surety that it feels constant. We come and go, but the oceans remain.

For thousands of years, our oceans rose and fell every day with the tide. Passing storms occasionally caused coastal damage, but the sea level itself did not change. The last time there was any kind of marked change in global sea levels was long before modern human civilisation developed. As our planet emerged from its most recent ice age around 20,000 years ago, the vast ice sheets on the northern continents broke up and melted away, and sea levels rose by about 120 metres, over the span of around 12,000 years. Then, for 8,000 years, global sea levels were static. Now, they are rising again – and they will be for some time. How long, and how fast, depends on us.

Sea levels started rising over two centuries ago, at the same time as the arrival of the Industrial Revolution and our increased output of greenhouse gases. At first, the rise was simply a product of the oceans expanding as they warmed. As time has gone on, though, the atmosphere and land surface have also warmed, and water from elsewhere has started to melt into the oceans, too – first it came from diminishing glaciers, then from major ice sheets. Since 2002, the Greenland and Antarctic ice sheets have lost a combined mass of over seven trillion tons of ice – about two-thirds of that from Greenland, and one-third from Antarctica. (And, while that's a staggering amount of ice, there's still over 25,000 trillion tons left frozen!) That melt since 2002 equates to about 20 millimetres of sea-level rise, spread out over the planet's oceans.

The rate of ice melt and sea-level rise has gradually picked up pace over time. A century ago, sea levels were rising around one millimetre per year; today they are rising between three and four millimetres per year, and that rate is set to get a lot higher as ice-sheet melt ramps up.

In the last 150 years, global sea levels have risen around 250 millimetres, or 25 centimetres. This may not sound like much, but that's without factoring in the shape of our shorelines. In most places, beaches are relatively flat with a slope between 1-in-20 and 1-in-100, meaning a small change in sea level translates to a large change in the high-tide mark over the slope of the beach. A 25-centimetre rise in average sea levels, for instance, would shift the high-tide mark 20 to 100 times further inland – somewhere between 5 and 25 metres. We are virtually guaranteed 50 centimetres of sea-level rise this century, meaning the high-tide mark is likely to shift inland between 10 and 50 metres. That means anything located within a few tens of metres of the current high-tide mark is going to be in trouble. If we allow ice-sheet melt to increase from there, we may see over one metre of sea-level rise by 2100, meaning that zone of trouble could extend up to 100 metres inland from the current high-tide mark, in some places at least.

Many of the world's major coastal and river-mouth cities are in danger. These places are home to nearly a billion people, with trillions of dollars of property at risk of inundation – everything from office and apartment buildings to roads, subways, ports and airports. One region where sea levels are rising faster than the global average is the eastern coast of North America, and things are set to continue that way through at least the next century. Miami Beach and dozens of other towns and cities along the eastern seaboard of the US already experience a lot of 'sunny-day flooding' where, even on a fine and windless day, a high tide floods the streets. Here in New Zealand, sunny-day flooding is also an issue. On a recent visit to Thames, I saw first-hand how a high tide on a fine day can submerge the roads near the waterfront.

The subways of Manhattan just escaped disaster when Hurricane Sandy brought a catastrophic storm surge to New York in 2012. With half a metre of sea-level rise, another super-storm like Sandy will completely flood the subways and inundate large parts of Manhattan Island. To prepare, city planners are looking at building extensive sea walls and green spaces around the margins of the island to protect people and property. All new buildings must be designed to allow the lowest two floors to be flooded.

At the same time as sea levels are rising, many major cities are also experiencing land subsidence – a gradual settling or sudden sinking of the Earth's surface. City authorities in Shanghai, China, have already built over 500 kilometres of sea walls and a series of gates to control river floods in order to try to protect the more than 26 million people who call the city home. Jakarta in Indonesia is in a similar situation, but on land that is subsiding more quickly because of both the weight of urban development and the water being pumped away from under the city centre. Massive sea walls are under construction and the city is planning to relocate 400,000 of its most at-risk residents.

London, the capital of the United Kingdom, might not be built on the coast, but it too is at increasing risk. The Thames, which flows through the heart of the city, is tidal all the way to London and beyond. Storm surges have flooded parts of the city many times in the past, providing the impetus in the 1980s for building the Thames Barrier, a series of gates erected across the river downstream from central London.

The gates are closed when very high tides or dangerous storm surges occur. When the barrier first went into operation, it was used on average once a year. Between 2011 and 2020, it was used on average over seven times a year. In the year from September 2013, it was closed 50 times – roughly once a week. Since the Thames Barrier was not designed to serve as protection against sea-level rise, the city is looking to spend billions on new protective infrastructure.

*

At the same time as sea levels are rising, the water in the world's oceans is also warming.

The good news? Since it takes a lot more energy to heat water than it does land, and since there's so much water on our planet, we don't need to worry about the oceans boiling any time soon. While our land surfaces have warmed on average nearly 2°C, the global ocean surface has warmed by less than 1°C. So, we can be grateful to our oceans for saving us from instant death: if all the warming we're causing went straight into the atmosphere, air temperatures would already have risen 40°C or more, and we would all be goners.

But – and here's the bad news – there are side-effects. Oceans might warm slowly, but they are also slow to respond.

They can't cool down again quickly, and changes to things like ocean currents and deep ocean circulation – which affect the distribution of nutrients in the water column and the health of marine ecosystems – take an exceptionally long time to reverse. Oceans also expand when warmed, and they lap at the edges of the Antarctic ice sheets. So, even if the climate stops warming tomorrow, we are still in for long-term changes in our seas.

There are many things living in the sea that are not used to temperatures changing, and are therefore very sensitive to any adjustment. For marine life, a small change in water temperature of even 1°C can have enormous consequences. It can cause mass migration of some fish species, and create damaging conditions for those species that aren't so mobile.

Then there are the reefs. The Great Barrier Reef off the coast of north-eastern Australia has already suffered several bleaching events in recent years because of very warm sea conditions and, whatever we do now, we are still going to lose about three quarters of it for good. At 1.5°C of global warming, around 20 per cent of it *may* survive, but if warming gets to 2°C it's likely the Great Barrier Reef – and other tropical coral reefs along with it – will die out completely.

At the same time as the oceans are warming, they are also becoming more acidic. Remember how I said they're soaking up the bulk of the heating the Earth is experiencing? Well, in the process, they are absorbing about a quarter of the carbon dioxide we're putting into the air. When carbon dioxide is absorbed into seawater, it turns into carbonic acid – so burning fossil fuels is like directing a pipeline of carbonic acid straight into our oceans.

I don't need to tell you that acid is generally bad news for all living things, but carbonic acid is especially nightmarish for creatures that use dissolved carbonate compounds to grow their shells or skeletons. This includes corals and many forms of plankton, micro-organisms and shellfish. As carbonic acid is added to the seawater, it becomes harder and harder for these creatures to get at the carbonates they need. A coral or a mussel has only so much energy to live its life, and it cannot devote too much of that energy to building its skeleton or its shell, so corners are cut and shells end up thinned or deformed. After a while, these creatures just can't grow properly at all. Of particular concern, several species of plankton rely on carbonates to grow their tiny shells. These creatures form the base of many food webs in the ocean, so if their populations crash that will put many other species higher up the food chain at risk, too.

If emissions keep increasing this century, ocean waters could become more acidic than they have been for around 20 million years. We know from the geological record that past mass-extinction events in the oceans have been associated with increased acidification. It's not clear where the threshold is for the next extinction event, but the more we acidify the oceans, the greater our chances of causing something very bad to happen.

I don't know about you, but I for one would rather not find out where the threshold for marine mass extinction actually is.

*

In the distant past, when the Earth was warmer, both poles warmed around twice as much as the global average. When it was cooler, both poles cooled about twice the global average. We know this from the geological record, but we've never seen it happen in real time.

The geological record gives us only snapshots of the past, in the form of averages over thousands of years. On that timescale, both poles have appeared to respond instantly. But now, we are getting to see warming happen day by day, and it's clear that the two poles do not respond in the same way at the same time at all. The Arctic has indeed already warmed at around twice the global average over the past century.

Meanwhile, the Antarctic has not warmed much at all. In fact, some parts are actually colder than they were 50 years ago.

This phenomenon is a product of differences in geography, the thickness of the ice and the state of the oceans. It's known as polar amplification, where any change in the Earth's net energy balance – because of, say, increased greenhouse gases – tends to produce a larger change in temperature near the poles than it does elsewhere.

The Arctic Ocean, which surrounds the North Pole, is stable and quiescent, and its sea ice is only a few metres thick. It is not subject to strong winds, so its waters do not mix much, meaning any surface warming that occurs tends to stay at the surface, melting the sea ice. This creates a feedback loop where, as the sea ice melts, more heat is absorbed into the water (because it's no longer being reflected away by the ice), and that causes more warming, which causes more ice to melt … and so on. This is why we've already seen nearly half the sea ice over the Arctic Ocean disappear in summer, and it's what will lead to the loss of the other half by the middle of the century.

The southern oceans, by contrast, are the most turbulent on Earth. Here, very strong winds blow over the seas, sucking warm surface water down into the depths. Antarctica's ice sheets are also thousands of metres thick, so there's an awful lot more to melt than there is at the North Pole. The south's ferocious oceans conspire with its enormous quantities of ice to make it a region where warming can only happen very slowly.

Given thousands of years, it is inevitable that the Antarctic will warm as much as the Arctic, just as it has done many times in the past, but this is really not something we want to see happen again. A large part of the Antarctic ice sheets would melt, raising sea levels tens of metres. If we can keep total warming well below 2°C, the chances are the Antarctic will remain mostly intact and the Arctic will cool down again, given time.

*

Back in 2003, Europe saw a heatwave over summer that is estimated to have been the hottest for at least 500 years – as far back as even the earliest records go – and it killed around 60,000 people. While day-time temperatures were often over 40°C, it was the extreme night-time temperatures that did the most damage. Most homes across normally temperate northern Europe do not have air-conditioning, and people could not cool down day or night. Residents, especially the elderly, essentially baked in their own homes.

In the two decades since then, heatwaves have continued to claim thousands of lives worldwide. A severe heatwave affected Europe again in the summer of 2019, breaking yet more all-time high temperature records, including France experiencing temperatures of more than 45°C for the first time in recorded history, and causing thousands of deaths across Europe as a whole. Most recently, the 2022 summer in the northern hemisphere brought record 40°C days to the United Kingdom, resulting in around 3,000 more deaths than usual in those aged over 65 in England and Wales – the highest figure since 2004.

A report compiled by over a hundred researchers and published in 2022 by reputable peer-reviewed British medical journal *The Lancet* gave a pretty confronting overview of the true impacts of heatwaves on a global scale:

Because of the rapidly increasing temperatures, vulnerable populations (adults older than 65 years, and children younger than one year of age) were exposed to 3.7 billion more heatwave days in 2021 than annually in 1986–2005, and heat-related deaths increased by 68% between 2000–04 and 2017–21.

High temperatures become especially difficult for our bodies to deal with when they're combined with high humidity.

This is because, when it's extremely humid, the sweat on our skin doesn't evaporate easily. Our bodies overheat more quickly, and that can lead to heat stroke. At humidity levels of 85 per cent or higher, temperatures in the mid-thirties are considered extremely dangerous for human health. When humidity is below 50 per cent, however, temperatures have to get into the forties to be considered extremely dangerous. But, once temperatures surpass 40°C, the conditions are classed as extremely dangerous no matter the humidity. As temperatures continue to climb, some hotter parts of the tropics and subtropics will see these sorts of 'extremely dangerous' conditions more often. Some may become literally uninhabitable before the end of this century.

Severe heat doesn't just cause significant problems for our bodies. It's an issue for our environment and our societies as a whole, highlighting just how interconnected our lives and the natural world are. That 2003 heatwave was associated with a drought across southern Europe that led to major reductions in crop yield. Same story in 2010, when another heatwave during the European summer brought drought conditions to large parts of Eurasia and North America, notably massive crop failures in Russia that led to wheat exports being banned. Several other major wheat-producing nations also experienced droughts and other natural disasters, leading to price spikes for basic foods such as bread. Those stresses are considered to have contributed to the unrest across the Mediterranean region that became known as the Arab Spring. It's just this kind of thing that leads defence forces to call climate change a 'threat multiplier' – already volatile places like Libya or Syria don't need much of a push to tip them into conflict and bloodshed, and food insecurity will certainly do it.

It goes without saying that things dry out faster in a warmer climate, so drought conditions can set in more quickly. Researchers and water managers now talk about 'flash droughts' – very dry conditions that develop over the space of days or weeks.

On top of that, some parts of the world are gradually becoming drier overall. Most of those places are already on the dry side, because they're on the edge of desert regions or sheltered from the rain by major mountain ranges. Of particular concern are the desert regions located in the subtropics – places like central Australia, southern Africa, the Sahara and the southwestern US. In these places, air flowing out of the tropics at high levels tends to sink and, as it does so, it warms and compresses. This leads to high pressures at the Earth's surface and to generally sunny skies, because the sinking and warming air causes clouds to evaporate. This means there's a semi-permanent belt of high pressures and sunny skies centred around 30 degrees of latitude in both hemispheres, because that's where these regions are located. This is known as the subtropical high pressure belt, and its latitude is moving slowly towards the poles as the warming climate causes tropical circulation to expand. As that happens, the regions on the poleward edge of the subtropical high are drying out.

One of the affected areas that is of most concern worldwide is the Mediterranean. Large numbers of people live around the Mediterranean coast, from North Africa to the Middle East and into southern Europe. In North Africa, the southern coast of the Mediterranean Sea is just beyond the Sahara Desert. As the subtropical high pressure belt inches towards the North Pole, the climate of the Sahara Desert is moving north, and is expected to eventually cross into the southern reaches of Europe, into Italy, Spain, Greece and the Middle East.

*

Residents of much of New Zealand woke on the first day of 2020 to the smell of smoke and, in the South Island in particular, skies an alarming shade of orange. But the source of the hazy air lay over 2,000 kilometres away in Australia. 'Waves of smoke are pouring across the Tasman from the Australia bushfires,' *Newshub* reported. 'MetService says smoke particles have been pushed over by an unbroken northwesterly that's flowing over the country like a river.'

If things looked apocalyptic to us, all this distance away, it was because the situation on the ground in south-eastern Australia was a kind of hell on Earth. Over Christmas 2019, wildfires blazed across nearly 200,000 square kilometres of the New South Wales and Victoria countryside, laying waste to communities and ecosystems alike. Over 3,000 homes were destroyed. Thirty-four people died in the fires themselves, and an estimated 445 more died as a result of smoke inhalation. An estimated one billion animals died, another two billion were affected, and an unknown number of species were driven to extinction. The carbon dioxide emissions from the few weeks during which the fires burned were equal to a whole year of Australian emissions from fossil-fuel burning.

The smoke plumes were clearly visible from space, and badly affected air quality across New South Wales, Victoria, the Australian Capital Territory and the city of Canberra. Smoke particles were deposited on snow and ice in New Zealand's Southern Alps, turning the glaciers orange and adding to ice and snow melt. The smoke made it all the way to South America before dissipating over the Atlantic and Indian Oceans.

It became known as the Black Summer, and it was the region's worst bushfire season on record – but such wildfires are only on the rise around the world. Later in 2020, in the northern summer in California and the western US, there was a repeat episode, echoing what had happened there in 2019, and the year before that. Record-breaking fires occurred in California in August, including the first 'gigafire', which burned more than one million acres (around 4,000 square kilometres). Five of the 20 largest fires on record in California occurred in 2020, making it the largest fire season on record in terms of area burned. The previous record year was 2018. In both eastern Australia and the western USA, the fires were preceded and made worse by drought conditions and extreme high temperatures.

In general, wildfires need three things: dry fuel in the form of trees and undergrowth, an ignition source, and weather conditions that aid the rapid spread of fire – things like high temperatures and strong winds. In places, climate change is contributing to all three. As an example, take California, where the area that's burned each year has gone up by a factor of five over the past 50 years. Here, dry fuel is increasingly available thanks to higher temperatures and drought; the hot, dry winds that blow from the interior of the continent often spread the California fires, and these winds are only becoming hotter and drier as the climate warms; and, in 2020, the ignition source for many fires came in the form of lightning strikes during intense thunderstorms associated with Tropical Storm Fausto. As ocean temperatures rise, tropical cyclones like Fausto only intensify, with the potential in the right conditions and the right locations to ignite more wildfires.

Australia's wildfires are similarly connected with climate change. The catastrophic Black Summer fires came towards the end of the hottest and driest year on record in Australia. The warming and drying in Australia have been clearly tied to the way the climate is changing, and both of those factors increase fire danger and fire intensity.

As the climate warms, we are also seeing fires occur in places where they have historically been rare. In the past few summers in the Northern Hemisphere, widespread fires have burned in Siberia, Alaska, the north of Norway and even Greenland. Places we normally associate with ice and snow are now apparently becoming lands of fire. Sites in Siberia near the coast of the Arctic Ocean have seen temperatures in the high thirties in recent summers, far beyond what is normally experienced in those places.

Arctic fires can become 'zombie fires', smouldering in peat through the winter only to flare up again in the following summer, leading to multi-year fires that can be hard to combat or contain. The vegetation on Arctic tundra has historically been fire-resistant, dense and moist, but as the climate warms the tundra is becoming much more fire-prone, especially inside the Arctic Circle north of 66 degrees north latitude. Fires are now penetrating regions of permafrost, frozen ground that holds a large store of methane and carbon dioxide that will be released at increasing rates as the warming continues. The soot from Arctic fires is falling on snow and ice in Greenland and over the Arctic Ocean, helping speed the melting of ice and adding further to the warming. The Arctic is the fastest-warming region on the globe, so in a way it's no surprise that we're seeing a big rise in the occurrence of fires there.

Someone once described climate change to me like so: 'It's getting warmer, and the cold bits are melting.' Well, as you can now see, that about sums it up.

4

Keeping track

My first job out of university, in the 1970s, was at the Meteorological Service in Wellington. At the time, there was a climatology division on the ground floor, where monthly records were tallied, and average monthly rainfalls and temperatures worked out. Everything was then filed away in the archives in the basement. To me, it all seemed very static and a bit boring – just adding up series of numbers and storing the answers – but I know now how ungenerous that impression was. Those climatologists were, in fact, doing vital work. By keeping track of just what the weather was up to from month to month, they – and others just like them all over the world – were building a store of information that's crucial to understanding how the climate is changing now. Without a good observational record, we wouldn't be able to understand very much at all.

Scientists have been recording observations of the weather and the climate for 150 years or more, at least in some places. Records of the surface temperature of both the land and the sea have been widespread enough to reliably estimate global averages since around 1880. Balloon-borne observations of the upper atmosphere became routine from the mid-twentieth century, spurred on by the Second World War and the need to support military aircraft operations. Satellites began providing comprehensive estimates of worldwide temperatures (and other things) in the late 1970s.

The oceans were only sporadically observed until the early 2000s, when the international Argo programme kicked off. Argo buoys cruise the planet's oceans 24/7, drifting with the currents and sampling the top 2,000 metres of the ocean column at various depths, before surfacing to transmit their data to a satellite. There are now about 3,000 Argo buoys drifting about out there, and New Zealand's own National Institute for Water and Atmospheric Research (NIWA) has deployed more than any other organisation in the world.

For two decades, it's also been possible to track the state of the world's big ice sheets using satellites that measure the gravitational field of the Earth. Since the ice sheets are so massive, they have their own gravitational pull, and as they melt that pull weakens a little. In the early 2000s, NASA's Gravity Recovery and Climate Experiment (GRACE) flew two spacecraft – dubbed Tom and Jerry – around the Earth in tandem to measure Earth's surface mass and water changes. The twin satellites kept in contact with a microwave range-finder, and when the Earth's gravitational pull caused even the most minute change in distance between them, the range-finder logged it. These tiny pushes and pulls – as small as the width of a human hair – were then converted into maps tracking the ups and downs of gravitational pull, which indicated changes in the mass of the ice sheets, changes in ocean-water mass, continental soil moisture and aquifer levels, even the flow of magma beneath the Earth's crust.

Several times a day, global weather centres take in as much information as they can from the land, the sea and the sky – amounting to hundreds of thousands of observations – and use it to build a picture of the state of the global atmosphere. This picture encompasses everything all over the globe from dozens of kilometres above our heads to ground level and down into the oceans. It's also the starting point of our daily weather forecasts, which are carried out using increasingly sophisticated weather-prediction models. The first such forecast was completed back in 1950, on a computer known as the ENIAC (Electronic Numerical Integrator and Computer) that was powered by valves and dials. This 'forecast' covered a 24-hour period for a single level of the atmosphere over North America only … and it took nearly 24 hours to produce!

Fully global weather analysis and forecasting by computer model got under way in a number of countries by the 1970s, and, here in New Zealand, first relied on punched cards and one of only very few computers in the country. This computer was located down the hill from the Meteorological Service at the Treasury, and in order to run weather-prediction experiments the meteorologists had to carry boxes of punched cards back and forth between the two sites. I joined the Meteorological Service at the same time as it acquired its first-ever computer system, so I didn't have to run boxes down to the Treasury, but I didn't escape the punched cards completely. One of my earliest jobs was setting up a system to predict tomorrow's maximum temperatures at main centres around the country, and the data for my first programs still had to be punched on to cards, using one of the machines in our data-entry room. Then, I had to use the punched-card reader to send the information I'd collected to what was dubbed 'the Trentham computer', at a government facility 15 kilometres away in the Hutt Valley. (This was well before the era of a computer on every desk!) Once the Trentham computer had done its job, the results would then be sent back to our line printer. This card-punch process lived on until the mid-1980s, when at last all the computing could be carried out on the Meteorological Service's own computers.

During the 1980s, my research turned to using statistics to predict aspects of the weather – day-time and night-time temperatures, winds, frosts, rain, even stuff like electricity demand in different parts of the country. It was very much my thing, and it had some pretty cool real-world applications, even getting me involved in the America's Cup in a small way. The 1986/87 Cup was held in Fremantle in Western Australia, a place that has a regular sea breeze most days. I got interested in seeing if I could predict when 'the Doctor' (as the sea breeze is known) would come in, and how strong it would be. After developing a system to make these predictions, I was asked to help out with writing code for use onboard the boat *KZ-7*, affectionately dubbed *Kiwi Magic*, during the America's Cup races. The software was set up to predict minute-by-minute changes in wind direction out on the course, using the latest observations from the mast. And that's how I ended up spending a few intense and fascinating weeks in Freemantle at the end of 1986!

That experience led to more with the yachting fraternity. Ahead of the 1989/90 Round the World race, Peter Blake and part of his crew came to do some meteorology training with us, and our research group developed route-planning software based on real-time global wind forecasts for their boat, *Steinlager 2*. The lead-up to the race itself was just a little bit exciting. On the day it was scheduled to start from Southampton in the UK – a Sunday here in New Zealand – I got a phone call very early in the morning from my colleague and team leader, Neil Gordon. 'The software has crashed!' he said. 'Nobody can work out what's going on. Can you go into the Met Service with me and run the same code with the same data?' There were only six hours before the starting gun went off. How could I say no?

At the office, we fired up the code and looked for problems. It took us hours, but we managed to locate the source of the error and build in a workaround, then email it to the UK just in time for it to be compiled onboard before the team set sail – and I'm glad to say Blake's team won the race! For those few hours in the Meteorological Service building, I felt like I was in a *Mission Impossible* movie – the clock ticking, the race's starting-gun imminent, the pressure on us to find the bug and fix the problem before it was too late. We didn't save the world, but it sure felt like it!

That kind of forecast model is now pretty standard stuff for many weather-sensitive businesses, including shipping, aviation and transportation. It's a level beyond what most of us need from our daily weather forecast, but many sectors of the economy have come to depend on timely and accurate weather forecast data – and, thankfully, the technology has come a very long way from those days in the late 1980s, when we had to hand-digitise faxed charts of forecast winds. We are now living in a golden age of weather and climate data. With the click of a mouse, we can learn about yesterday's global ocean temperature patterns, the distribution of sea ice at both poles, global weather patterns, the state of the El Niño/La Niña cycle, the amount of carbon dioxide in the air, and much more. Every month, major analysis centres report on what global temperatures have been in the past month. We know very accurately exactly what is happening with weather and climate all over the world. For me, and for the science community generally, this is a wonderful boon to research and to understanding.

*

Mathematical models based on physics have been around and have been used for predictions for a long time – and one of the first and most widely celebrated examples came in the form of a comet burning across the night sky towards the end of 1758.

Over half a century earlier, English astronomer Edmond Halley had published calculations showing that comets observed in 1531, 1607 and 1682 had very similar orbits – so similar, in fact, that Halley believed them to be just the one comet, returning periodically. He predicted this comet would return in 1758, and so it did. Halley was no longer alive to see it, but the comet was fittingly bestowed with his name. (And, actually, there's reference in the Talmud to 'a star that appears once in 70 years and makes the captains of ships err', so it's possible humanity already knew of this periodic comet millennia earlier.)

The confirmed return of Halley's comet was a pivotal moment. It showed how our knowledge of past events can be put to use predicting what lies ahead. And, although we might have more data and technology at our disposal than Halley did, our process when it comes to weather and climate science remains essentially the same: we look at the past to predict the future, based on the laws of physics. There's just one thing we have to do first, and that's build as accurate a picture of the physics of the climate system as we can. For this, we use climate models.

Climate models work on the same principles as weather-prediction models: they use computer code to simulate how the laws of fluid motion and heat transfer will shape temperatures, winds, air pressures and so on. However, while weather-prediction models aim to determine as exactly as possible how things will evolve over the coming days, climate models aim to see how things will evolve over years or centuries. That means focusing on the slower-changing parts of the climate – especially the oceans, but also the land surface and the global distribution of ice and snow. The atmosphere is important, too, but getting the starting conditions right is not as much of a concern as it is with a weather-prediction model. To predict the sequence of weather accurately over the next few days, it's important to begin from the most accurate starting conditions. By contrast, to predict the climate, we are looking at averages and statistics over decades or longer, not the day-to-day sequence of the weather, so the starting conditions don't matter much. It's the boundary conditions we care about – things like the brightness of the sun, the amount of greenhouse gases in the air, the state of the ice sheets.

Most of the recent advances we've seen in weather observations and other technologies have been primarily in service of improving weather forecasts: the better we can specify the state of the weather today, the better it can be predicted for tomorrow. Of course, the enormous amount of data gathered can be applied to tracking longer-term climate trends, but it's not as straightforward as just using the raw information. Whenever a new set of observations has come on-stream, it has changed the look of the weather analysis, because we can suddenly see things that were hidden before. Whenever a new weather-prediction model has been introduced, the weather analysis also changes, because we use the model to compare with the observations. What we observe has to line up with the physics represented in the model.

Changes in models or in available observations can sometimes show changes that actually have nothing to do with climate change – sequences of temperatures or rainfalls or whatever can change simply because we observe or model them better. To get around this, there has recently been a huge global effort to essentially re-do daily weather analyses, from yesterday backwards, to the 1970s, 1960s, the 1950s and even earlier, using the very best of today's weather analysis and forecasting models, and the billions of observations that have built up over time. It's a staggering effort, and takes years to complete, even on the fastest supercomputers – but the results so far have been worth it. These re-done analyses and forecasts have greatly advanced our understanding of how the weather and the climate vary, and how the climate has changed since the mid-twentieth century.

Going even further back, we can also use climate models to simulate any past climate – we can travel millions of years back in time to follow the peaks and troughs of the ice ages, or to re-create the hothouse climate of the Eocene, when New Zealand was both a different shape and located much closer to Antarctica. To do so, we need to specify the brightness of the sun and the amount of greenhouse gas in the air, and if we go back in time far enough we also need to worry about the location and shape of the continents, as these have changed over millions of years.

In a climate model, the Earth essentially becomes a big jigsaw puzzle. We can put the continents wherever we like. We can add information from geological records of the distant past and apply the findings of lab experiments measuring how different gases absorb energy. We can include theoretical studies of how the fluid atmosphere and the oceans move, and factor in the implications of temperatures gathered from both the ground and the sky far above our heads. We can account for ecosystems and their role in the carbon cycle. If the information exists and it affects the climate, we can include it. In this sense, climate modelling is a means of using physics and chemistry to translate what the past tells us into an understanding of the future.

One of the main ways we use climate models is to ask 'What if?' What if we doubled the amount of carbon dioxide in the air? What if a major volcanic eruption happened? What if Australia disappeared? What if the Antarctic ice sheets were only a metre thick? Those last two might seem pointless – we all know Australia *is* there, and the Antarctic ice sheets are *thousands* of metres thick – but they're not silly questions. If we simulate the climate both with and without the continent of Australia, for instance, we can work out what effect Australia has on things like atmospheric circulation, tropical rainfall, jet streams and so on. Same story with the wafer-thin ice sheets – they can tell us how much effect the real ice sheets have on the winds over the southern oceans.

That's the great power of models. We can build simulations of the Earth system that are good enough to reproduce all the main features of the climate: day-to-day weather, seasonal variability, El Niño events, the location of the monsoon regions and the deserts, the large-scale circulations of the oceans and atmosphere. These models look a lot like the real Earth, but they enable us to make changes that are impossible to do on the real Earth. We have no second planet where we can adjust the composition of the atmosphere, or the arrangement of the continents, just to see what happens. But, in a model, we can do these things, and it is a very powerful way to tease out cause and effect, to understand why the climate looks like it does and why it is changing the way it is.

To carry out climate change experiments with climate models, we need to run our models both with and without greenhouse gas increase, over the same period of time. We might start the model run at the beginning of the twentieth century and let it go through to the end of this century or beyond. In the run *with* greenhouse gas increase, we tell the model how bright the sun is each year, how much greenhouse gas is in the air, and when major volcanic eruptions happened. For the past century or so, we already know exactly what happened when. For the future, we need to specify a scenario – for greenhouse gas levels, for how bright the sun will be, and for when major volcanoes will erupt. For sunlight, it's standard practice to assume it will be constant or will follow the typical 11-year sunspot cycle into the future – but it is also possible to vary that, and look at the effects of, say, a major decrease in solar output. Since volcanic eruptions are unpredictable, the simplest and most reasonable thing to do is say there won't be any, as volcanoes affect temperatures for only a year or two.

Climate modelling informs, at least in part, the IPCC's regular assessments on global warming and the state of our climate. Since the IPCC was formed back in 1988, our climate models have become remarkably accurate, and much more sophisticated, but their message hasn't changed radically in more than two decades: the more greenhouse gases we emit, the greater the damage. The most recent assessment was sobering. Many of the latest models showed faster warming with increased greenhouse gases than earlier models had shown. This seems to be related to improved handling of clouds in the models. Some of the most complex models are suggesting that a warmer climate may have less clouds, so less reflection of sunlight, so more and faster warming. However, the jury remains out on this, as we don't have observations to confirm those model results, and our records of past climate changes don't really support the size of temperature changes simulated. Still, such results are concerning, and suggest we have even less time to reduce emissions if we wish to stop the warming at 1.5°C or even at 2°C above pre-industrial levels.

Thanks to climate models, we have a very detailed picture of both how the climate has changed in the past and how it is set to change in the future, uncertainties aside. We know what's happening, why it's happening, and what will happen in future if we carry on as we have done in the past. Science defines the problem, but human activity controls it.

*

For all of the extreme events already occurring in our world's climate – from heatwaves to hurricanes – one question is of paramount importance: What's the human fingerprint? Or, to put that another way: Did human-driven climate change make this extreme event more likely to occur? And, if so, how much worse or more intense was it because of climate change?

These questions are very different from asking whether climate change *caused* an extreme event, and to answer them, we can once again turn to climate models. This is the realm of climate change attribution science: the quest to determine which events are attributable to the changes in our climate that we have already caused to occur, and which events are not. We find that not all extremes show a human fingerprint, so far – but, as the scale of climate change increases, those human fingerprints will become more and more obvious. It's an area of research that grew out of the effort to more generally understand how an increase in greenhouse gases is changing our weather and our climate.

Researchers can determine, for instance, how much temperature rise is attributable to human-driven climate change by running climate models that cover the past century, both with and without greenhouse gas increase. The results show that, without greenhouse gas increase, global temperatures would have actually gone down slightly over the past 70 years. They also show that, without humans burning coal and oil, the increase in global average temperatures since the 1950s simply would not have happened. The rising temperatures are clearly down to us.

That kind of broad-scale attribution is fairly straightforward, and confirms what we already know. But what about homing in on a particular heatwave, or on an extreme event that happened in a particular place on a particular day? The trick is to look at how often a climate model throws up an extreme occurrence similar to the one observed, at about the same time of year and in the same place, with and without greenhouse gas increase. Let's say a heatwave hits Wellington, and on 23 January the day-time temperature is 10°C above the average for that time of year. Using climate models to simulate the climate with and without greenhouse gas increase, we can count the number of times that the temperature near Wellington got to be 10°C above the model's average in January for the years around the year the extreme actually occurred. The models might show that, in the simulations with greenhouse gas increase included, those high temperatures occurred three or four times more often than they did in the simulations with no greenhouse gas increase. That would then allow us to say that human-caused climate change made the temperature extreme three or four times more likely than it would have been otherwise.

Thanks to attribution research, we know that the heatwaves different regions of the world have experienced this century have a large fingerprint of human-caused warming. Heatwave conditions seen across Europe in 2019 are at least ten times more likely now than they would have been in a world with no greenhouse warming. The Siberian heatwave that occurred in 2020 and saw temperatures reach almost 40°C would have been almost impossible without the warming of the past century. Closer to home, the bleaching event on the Great Barrier Reef in 2016, caused by very warm seas, is estimated to be nearly 200 times more likely now than it would have been without global warming. And the marine heatwave that affected the Tasman Sea and New Zealand region in the summer of 2017–18 was found to be virtually impossible without the warming of the past century. The very warm seas observed that summer just did not happen in any of the model runs that didn't include the global warming brought about by greenhouse gas increase.

*

My old friend, Dr Sam Dean, is one of New Zealand's leading climate change experts. His official title at NIWA is Principal Scientist – Climate, and in recent years he's shifted away from using climate models to ask 'what if' to instead focus more closely on attribution research. The way he sees it, modelling the future might be scientifically interesting, but people have become desensitised to the sorts of doom-and-gloom scenarios the models can illustrate. Attribution research has much more of a chance of making an impact. If we can see, in real time, what the effects of greenhouse gas emissions are – and, potently, put a dollar figure to the damage we're causing – he, and other scientists like him, hope we might be spurred to make the changes we need to.

After the Far North of New Zealand was hit by a ferocious storm in July 2014, Sam and his colleagues used climate models to determine what role human-induced warming played in the floods that ensued. For perspective, the town of Kaikohe usually receives around 180 millimetres of rainfall in July, but during the 2014 storm it was deluged under 477 millimetres – and other parts of the region were even wetter. Farms across Northland were left underwater and homes uninhabitable. By comparing observed data with models simulating the region's usual conditions, Sam and his team were able to say with 'medium confidence' that human-induced climate change accounted for 47 per cent of the risk of that extreme weather event happening. In a paper published the following year, they noted that an event as extreme as that of July 2014 is estimated to change from an approximately 1-in-350-year occurrence in a world without greenhouse gas increase to a 1-in-200-year event with human-induced effects on the climate. The same analysis will be carried out on the Auckland floods of January 2023, and no doubt a human fingerprint will be found.

It's this sort of analysis that can be used to put a price on the impacts of climate change. For instance, in the decade to 2021, NIWA estimated that extreme weather events such as floods, heatwaves and droughts collectively cost New Zealand $940 million. That really puts the impacts of climate change in perspective. But, in order to truly communicate the immediacy of the impacts of human-drive climate change, it's imperative to share what we know as soon after an event has occurred as possible. In 2015, the World Weather Attribution (WWA) group was set up with precisely this aim. A thoroughly international initiative, the WWA brings together climate scientists and specialists from the UK, the Netherlands, France, the US, Switzerland and India to provide robust and speedy assessments on the role of climate change in the aftermath of an extreme event.

There's also a pretty neat citizen science aspect to attribution research. Through the website climateprediction.net, anyone can sign up to participate in what's touted as 'the world's largest climate modelling experiment for the twenty-first century'. The project, based at the University of Oxford and overseen by a team of climate scientists, computing experts and graduate students, allows people to run climate models on their home computers to map how climate change is affecting our world. Thanks to the involvement of thousands of volunteers, the project is able to run simulations that focus on extremes in several specific areas of the globe. One such region is Australia and New Zealand, and our model simulations are managed by the climate research group at NIWA, in collaboration with Melbourne University. There's great power in a project like this: the more people who get involved, the more model simulations the project can run, and the more detailed the statistics can be.

Computer power is a major factor in climate modelling, and in 2018 NIWA's High Performance Computing Facility commissioned three interconnected supercomputers to increase our modelling and forecasting abilities. Two of these powerful computers are located in Wellington and one is in Auckland, and together they can process more than two thousand trillion calculations per second. As well as climate modelling, they lead investigations into forecasting weather-related hazards, tracking our freshwater resources and understanding the systems driving our oceans. They allow us to run models at much higher resolution, with fewer approximations made – meaning, essentially, we can effectively break the Earth down into smaller boxes to make more finely tuned calculations and estimates. It's quite a step forward from ferrying boxes of punched cards to and from remote computer centres!

Supercomputers also open the door to using artificial intelligence in modelling and climate science. Using machine learning methods, we can feed historical climate data into the computers to simulate Earth's processes. For instance, rather than prescribing a certain amount of ozone in a model, we can instead use a computer algorithm to generate the ozone layer. Therein lies the promise of AI: not only can it improve what we already know, but it has the potential to produce new knowledge, to come up with insights we haven't thought of.

To me, to my colleagues like Sam, and to others like us, there's beauty to be found in climate modelling. These models uncover what was once hidden in the past, and they show us what's possible in the future – good or bad. It's impossible not to appreciate the artistry inherent in these acts of coding, converting our knowledge of science into useful pieces of software. In fact, when the UK Met Office had its new headquarters built in Exeter, they commissioned beautiful glass panels cast with lines of code from their climate model, a lovely melding of art and science. In a way, by re-creating our world in various forms, the models serve as a reminder of the true wonder of the source material. Our planet, with its weather and its climate, its oceans and its atmosphere, is a work of art. As Sam himself succinctly puts it, 'The Earth is beautiful. The clouds are beautiful.' I myself find looking at the clouds, the sky, the ocean endlessly restoring.

Behind all the data and the predictions and the models, it's important not to lose sight of this fact, because it's this miraculous planet – and our ability to live on it – that we're all fighting so hard to protect.

II

THE LOCAL

5

Home turf

Global warming and climate change are big things. They affect the whole planet – every society and every community in the world. The scale is so immense that it can, for many of us, be difficult to truly comprehend. Something so big is overwhelming. It's nearly impossible for our brains to process.

To grapple with the global issue of climate change, it's helpful to home in on the local. For those of us here in New Zealand, that means doing more than paying attention to terrible events unfolding in faraway lands. Understanding climate change is about more than the fires and droughts and heatwaves hitting Europe and Asia and the USA. It's about more than the ice melting in Antarctica and

Greenland. It's about our own backyard. Because, just like the rest of the world, we too are feeling the effects of climate change. Increasingly intense extreme weather events are damaging the things we rely on – our homes, our roads, our livelihoods – and affecting every one of us personally.

Let me ask you, how has your life been touched by climate change? Perhaps your region recently experienced an unprecedented flood, bigger and wetter than anything ever seen before? Maybe it changed the course of your favourite river, or eroded a street you used every day, or even flooded your home?

Or maybe you live by the coast, and intense storms have eroded land once deemed safely distant from the water? Perhaps you have watched as the waves forced their way further inland than ever before, on the back of higher seas?

If you live in a city, you may have been subjected to stinking-hot days that buckle the train tracks and hamper your work commute? Equally possible, it might be that heavy downpours have stymied your transit by causing tracts of land to slip down over the roads?

Or maybe things do actually look different when you peer out the window? Are the usually dun-coloured hills greener than ever? Is there less snow on the mountain peaks? Are the trees and plants happily growing in your garden now ones that were absent when you were a child?

All of these things are real. We see them with our own eyes. We live through them. Our home turf is changing as the global climate shifts. Coping with this change is challenging, because it means accepting that the land we grew up with is gone. Everything is different now.

*

I worked at NIWA during the nineties, and we'd often have international scientists come to do research with us. I'll never forget a Japanese colleague's reply when he was asked how he'd found his year in New Zealand. 'Well,' he informed us, 'it's always springtime here!'

It was a pretty apt summation of our climate. Things are indeed fairly benign here: not too hot, not too cold and often quite windy. New Zealand's climate is classed 'temperate maritime', meaning it's not very extreme and is influenced by the oceans and the winds off the oceans. Broadly speaking, our days are a mix of sunny and dry, then wet and windy. Over the country as a whole, there is generally plentiful rainfall, and most food crops will grow somewhere here.

There are a couple of main ingredients in New Zealand's climate soup. The first: the westerly winds that circulate around the middle latitudes of the Southern Hemisphere. At

97

all times of year, these winds blow across the country, and particularly the South Island. The second ingredient is our topography, notably the Southern Alps in the South Island, and the Tararua and Ruahine ranges in the North Island. These two elements – the westerlies and the land – conspire to determine our average climate.

Generally speaking, when the moist westerly winds strike land and ascend over the mountains, they make our western coasts our wettest. The eastern regions, meanwhile, are sheltered on the other side of the main mountain ranges and therefore remain much drier. If, one year, the westerlies are stronger, we'll end up with even wetter conditions in the west and drier in the east; weaker westerlies will lead to the reverse. And, if the westerlies are more from the southwest, temperatures over the whole country are generally lower; but, if they are more from the northwest, temperatures tend to be higher.

Being a truly mid-latitude country – the halfway point between the Equator and the South Pole (45 degrees south) passes through Oamaru and Queenstown – New Zealand is exposed to influences from the tropics and subtropics to the north, and to influences from the Southern Oceans and Antarctica to the south. But, although 'temperate maritime' is a good description of our climate generally, it hides a lot of regional variability. While the northern tip of our stretched-out country very nearly reaches the subtropics, our southernmost point projects into the surly Southern Ocean.

Places on the West Coast of the South Island are some of the wettest on earth, right up there with the monsoon regions of India and South Asia. The global average rainfall is around one metre per year, but somewhere like Milford Sound receives over six metres of rain in an average year, while elevated locations west of the main divide record up to ten metres – as in 10,000 millimetres! – or more of rain or snow on average each year. The upper half of the North Island, at the mercy of subtropical weather influences, ends up reasonably well watered by passing subtropical storms.

The closest New Zealand gets to a continental climate is in Central Otago, which sits in the rain-shadow of the Southern Alps. Annual rainfalls here tend to be around half a metre per year – only about a tenth of those west of the Alps, just 50 kilometres away.

Our annual average temperatures also vary depending where in the country you are. In Northland, the average each year is usually around 17 or 18°C, while it's only 9 or 10°C in the far south. Our summers are generally mild, but extreme temperatures in the high thirties can occur in the east and in Central Otago. Winters, too, tend to be mild, apart from inland Canterbury and Otago, where the frosts can be severe. The mountainous areas also, obviously, have much colder conditions.

Specific climate patterns also change things from year to year. An El Niño summer, for instance, is often colder and windier than normal, as the sunny and dry days withdraw to the north, while the stormy westerly winds from the Southern Ocean hammer most of the country. La Niña, on the other hand, does the opposite, drawing high pressures and the sunny skies to most of New Zealand, while the winds and storms head south towards the Antarctic coast.

Being open to weather influences from the tropics, the Southern Oceans and Antarctica also has an effect on our climate. The Southern Annular Mode (SAM), for instance, is a ring of climate variability encircling the South Pole that extends to our environs, and brings alternating windiness and storms to the mid-latitudes.

There's a lot about New Zealand that makes it appealing to those interested in studying climates of the past: our location, the interplay of the westerly winds and the mountains, our exposure to a wide range of climate influences. And, just as the story of the world's past climate can be found in the ice or on the sea floor, it's possible to read New Zealand's history in cores taken from our land. Back in 2016, a group of researchers drilled deep into the bed of Lake Ōhau, in the South Island, to remove two sediment cores that told of weather events, vegetation changes and human impacts in the period back to some 18,000 years ago. As well as signalling the speedy (in geological terms) retreat of glaciers from the area, these cores also traced the impact of the westerly winds.

'Lake Ōhau is right at the edge, at the northern boundary of these westerly winds ... and the westerlies control a lot of our weather, and ultimately our climate,' project leader Richard Levy, from GNS Science, told *RNZ* at the time. 'What we're interested in knowing is how that wind system has evolved through time, and moved or responded to large-scale features in the global climate system.'

In other words, the way the westerlies have interacted with the land, and with influences like El Niño and the SAM, over the centuries tells us a lot about the natural climate changes that have occurred in our country's past. It also tells us how these things work together today, and might conspire in the future.

*

On the morning of 7 February 1973, I biked to school. At eight-thirty, it was already stinking hot and the asphalt on the road had started to melt. When I got to my classroom, the teacher let us remove our ties – pretty radical, at the time – and within a few hours the temperature had surpassed 37°C and we were all sent home. My friends and I, along with the entire city of Christchurch and its surrounding area, headed for the nearest beach, pool or garden hose. Anything to help us cool down.

Rangiora, just under 30 kilometres north of central Christchurch, hit a high temperature of 42.4°C that day. It was – and remains – the hottest day ever recorded in New Zealand, and was a direct product of our country's geography and the winds blowing over it. Everything aligned just so: the nor'wester was exactly at right angles to the Southern Alps, and the air that went up and over the mountains was very moist, so a lot of rain fell on the West Coast, which heated the air. Then, when that air came down the other side of the alps, it compressed and became still warmer at sea level.

Westerly winds blow across the Tasman Sea towards New Zealand year-round. Since the sea's surface temperature changes only slowly, the on-shore winds keep land temperatures pretty even too – meaning that the wet western parts of the country tend to show little variation in temperature, outside the seasonal cycle. For our eastern regions, however, it's a different story. In order for those westerlies to make it to regions like Canterbury, Hawke's Bay or Gisborne, they have to get up and over the massive mountain ranges bisecting both islands.

In the South Island, for instance, the moist air that strikes the West Coast must rise to travel over the Alps. As this air rises, it cools, and quickly becomes saturated with water vapour, since cooler air holds less water vapour. As the water vapour condenses back into liquid water drops, it releases the energy it took to evaporate that water in the first place. This is what is known as 'latent heat' – heat energy that is locked up in water vapour until it condenses again. All of the latent heat released when clouds and rain form west of the Southern Alps is released into the air, warming it up. Then, when the air makes it to the east of the Alps and starts to descend again, it compresses and warms further. As a result, the winds that blow down onto the Canterbury Plains are warmer than they were when they first reached the West Coast. If enough moisture is released in the west, and if the descent to the plains is strong enough, the downslope warming can be extreme – just as it was in February 1973.

Just how warm it gets east of the mountains depends on a number of factors. The way temperatures vary in the vertical is important, as is the exact angle of the wind when it comes into contact with the South Island. The nearer the wind flow is to being at right angles to the line of the Southern Alps, the more likely the downslope warming will be strong.

When this sort of natural variability is combined with human-driven warming, it leads to even more extreme extremes. Northwest winds have always been warm in Canterbury – so, add a warming trend, and it becomes even easier over time for a northwesterly day to reach record high temperatures. Given that average temperatures are rising, and the amount of moisture (and hence latent heat) in the air is also increasing, it only seems inevitable that the 1973 record will be beaten, sooner or later.

*

Growing up in Canterbury, the nor'wester was a real feature of life. It brought hot, dry days and often gusty winds. And, while it could sometimes be really unpleasant, in the right conditions it made for perfect kite-flying, when the wind was steady and not too strong, the temperature warm but not too hard to bear.

My dad taught me the basics of kite construction: two sticks lashed together into a cross, a string run around the points, everything covered in a layer of paper, and – the final touch – a tail attached to give the kite some stability. The best tails were made with old pantyhose, which were light and drapey but big enough to do the job of getting the kite to stay upright. Most of my kites were pretty simple affairs, but one especially robust model used larger bamboo poles and was covered with an old bed sheet. It was heavy, but a good nor'wester got it aloft no trouble.

There was a big paddock across from our house that didn't seem to be used for anything much, so on those perfect nor'west days I would race across the road and get my kite flying way up high on a few hundred metres of string. Then, I would tie the string to a fence and leave it there. I'd pop home for lunch and return to find my kite still cruising. It would stay up there for hours. Now I know that's because the winds well above ground level were stronger and more reliable than those down on the ground.

On one memorable occasion, I took one of those cheap little snap-together balsa-wood aeroplanes that were popular back in the day outside in a nor'wester. I must have chosen a good day and a good moment to launch it, because the wind just picked it up and carried it east at speed. I jumped on my bike and followed it for as long I could, but it had soon disappeared from sight, headed for Christchurch.

Quite often, the nor'wester would be so strong down at ground level that it made biking – my main mode of transport at the time – difficult. On the morning of 1 August 1975, I jumped on my bike to get to the cleaning job I had at a gym near the university. It was fairly windy, but I got there OK. When I came outside a couple of hours later, however, I found the ten-metre-high steel posts holding up the gym's sign had been bent flat to the ground. Somehow, I managed to get back home (still on my bike, as I recall), only to discover the shed I kept my bike in had become a pile of sticks. Thankfully, our house was OK, but over the next few hours we all nervously watched the glass in the big picture windows flexing in the gusts.

That day, the northwesterly winds gusted up to 172 kilometres per hour at Christchurch Airport, and similar gusts were recorded from Kaikōura to Timaru. Many homes and businesses were damaged, with some buildings demolished, and power outages were rife, with lines down across the region. Forests across Canterbury suffered severe damage, and my job the following summer ended up being one of the team helping to clear wind-blown pine trees off a beach north of Christchurch – the trees had been lying there since August. All across Canterbury, similar scenes played out, and recovery from the devastation took years.

Those extreme gusts had a lot to do with how the air descends from the tops of the Southern Alps. Not only does it bring warming, but it can also increase wind speeds. High up in the atmosphere, winds tend to be stronger – at jet-stream level, around 10 kilometres up, for instance, they can reach as high as 200 kilometres per hour. The air that descends from the tops of the Southern Alps – winds blowing at three or four kilometres altitude – can be travelling at 100 kilometres per hour or more, and sometimes those speeds make it to ground level as the nor'wester touches down in Canterbury. So, just as the nor'wester contributed to the perfect conditions for the warming that gave us the 1973 scorcher, conditions can also be just right for bringing really fast-moving air down to ground level.

*

New Zealand is occasionally affected by ex-tropical cyclones – storms that used to be tropical cyclones, but changed on their way to us. That change is what's called extra-tropical transition, and it usually happens around 25 degrees south latitude. A storm poised on the edge of the subtropics, about to lose its tropical nature, is a little like a pinball perched at the top of the track, about to head down into the bumpers. Tiny changes in the speed of movement, the winds and other environmental conditions can launch the cyclone in a slightly different direction. Most cyclones pass well to the east of New Zealand, but every now and then they'll track close enough to be felt somewhere in the country, usually the northern part of the North Island – and sometimes they come very close to our coasts, or even run right across the country.

Ex-tropical cyclone Giselle, the storm that ran right across the North Island and passed just east of the Wairarapa coast on 10 April 1968, is one of the most infamous to have hit our shores in recent history. It started life in early April near the Solomon Islands, and tracked just to the west of New Caledonia before beginning its transformation into a powerful mid-latitude storm. Its location when it reached our shores was near perfect for funnelling hurricane-force southerly winds and phenomenal seas into Cook Strait and Wellington harbour. The fury of the storm was short-lived, but in the few hours during which it assaulted Wellington it managed to catch the interisland ferry *Wahine* at the mouth of the harbour. The ferry went aground on Barrett Reef at the harbour entrance and came to rest on its side just off the beach at Seatoun. All up, 53 people lost their lives in the disaster. The ship was only metres from the beach, but the winds were so strong that most passengers, in lifeboats or in the water, were blown right across the other side of the harbour, coming ashore around Eastbourne and Days Bay.

Giselle was one of the most devastating ex-tropical cyclones to hit New Zealand, but hardly the only one. Just 32 years earlier, in 1936, a similar but nameless storm caused phenomenal damage across the whole North Island and almost sank the interisland ferry of the day. And, two decades after Giselle, in March 1988, Cyclone Bola lashed the east coast of the North Island and devastated properties and farmland in Gisborne and Hawke's Bay. In 2018, ex-cyclones Fehi and Gita, only weeks apart, both pounded New Zealand, bringing strong winds, torrential rain and storm surges to parts of the South Island. In January 2003, ex-cyclone Hale brought torrential rain with slips and flooding to much of the eastern North Island, especially between East Cape and Gisborne.

In February 2023, ex-Tropical Cyclone Gabrielle visited New Zealand, running down the east coast of the North Island. The damage, from extremely heavy rainfall and to a lesser extent from strong winds, was very dramatic. Homes, farms, orchards, roads and other infrastructure were badly damaged or destroyed in Northland, the Coromandel, and down the east coast of the North Island from East Cape southwards. At the time of writing, 11 people are known to have lost their lives and some communities remain cut off from road access. It is clear that Gabrielle was one of the worst storms to hit New Zealand and we can be sure that some of its ferocity is a result of global warming and climate change, notably as it formed and tracked over very warm ocean waters. Once the scientific analyses are completed, we'll know just how big a role climate change played, but already we know the central low pressure of the storm (a measure of intensity) was lower than either Bola or Giselle (the Wahine storm). That tells us the storm was unusually intense, probably as a result of the warmer atmosphere and sea surface. We also know that atmospheric moisture levels are now between 3 and 5 percent higher than at the time of the earlier storms, no doubt contributing to the deluge, and no doubt a result of the warming climate.

*

Another windy aspect that brings big fluctuations in New Zealand's weather is the Southern Annular Mode (SAM) – but, although it was first identified in the 1970s, it was largely ignored for many years.

Basically, the SAM is there because the storms and the westerlies talk to each other. Most of the time, the core of the storm track and the strongest westerlies sit south of New Zealand, roughly at the latitude of Campbell Island. Storms tend to form where the winds are strongest, and as they grow and move, they help to keep those westerlies strong. Sometimes, a storm will start to form somewhere north of 'normal', perhaps near New Zealand, and that drags the strongest winds north as well. There's no particular reason this happens – it's just the way the circulation of the atmosphere works. Anyway, since everything is whizzing around at such high speeds, this northward wiggle is quickly translated around the rest of the Southern Hemisphere. Should a storm instead form to the south, it's the same story, just the other way around: the westerlies dip south. The net result is that the westerly winds and their attendant parade of storms move north and south every few weeks, concertinaing in and out from the Antarctic coast.

When the SAM moves south – in what is known as its positive phase – storm activity picks up near Antarctica, and the westerlies ramp up over the Southern Oceans near 60 degrees south latitude, not far from the Antarctic coast. Meanwhile, high-pressure systems (anticyclones) move in over New Zealand, bringing settled weather and warmer conditions both here and in a ring around the hemisphere in the middle latitudes.

When the SAM moves north – in what is known as its negative phase – storminess decreases near Antarctica and winds become lighter near the Antarctic coast. Over New Zealand, the westerly winds pick up and the weather becomes cooler and stormier.

Thanks to the SAM, New Zealand can go from settled and sunny to windy and stormy every few weeks or so. And, while South America and Tasmania also feel the effects of the SAM, it turns out that we're in just the right spot to get the maximum hit: we experience the biggest decreases and increases in storminess during the different phases of the SAM of any location around the middle of the Southern Hemisphere.

The flipping back and forward from positive to negative happens essentially at random, as it comes down to the tracks of individual storms bouncing around over the Southern Oceans – and this is why the SAM was ignored for so long. It wasn't very useful for weather forecasting, and seemed merely an interesting curiosity – at least until a long-term trend in its behaviour started to reveal itself. Over time, it turned out that the SAM had been gradually becoming more positive:

more months in the 2000s saw the positive SAM, while most months in the 1970s saw the negative. This gradual change has affected the climate of most countries in the Southern Hemisphere, especially New Zealand.

Another thing that emerged around the same time was that the SAM trend doesn't actually have all that much to do with global warming (though it certainly helped). Instead, it is mostly a by-product of the ozone hole. When ozone is removed from the atmosphere, temperatures drop, because ozone is one of the few gases in the air that absorb sunlight, and that's exactly what's happened in the Antarctic stratosphere every spring since the 1970s. Colder air over the pole means a larger north–south temperature difference, so stronger westerly winds in the stratosphere – and that, in turn, has translated to stronger westerlies over the Southern Oceans, with more weeks and months of the positive SAM. The consequences for New Zealand have been more months of clear skies and lighter winds than we might have expected if the ozone hole wasn't there.

The thing is, the ozone hole won't be there forever, thankfully. Once it was discovered in the 1980s, first the science community and then the policy community worked frantically to understand what was happening and how to fix it. The result was the 1989 Montreal Protocol and the ban on chlorofluorocarbons (the gases implicated in the destruction of ozone). Now, the ozone hole is expected to be completely healed in 50 years or so.

Great news – and a great example of just how much impact human activity can have on our environment and our weather, for ill or good.

6

On our doorstep

In many ways, New Zealand is a climatically 'lucky' country, but even we are feeling the effects of climate change. The list of changes we're seeing in our local environment is a direct echo of what's happening in the rest of the world: heatwaves, drought, floods, fires and winds. Basically, more extreme everything. Our ice and snow is melting. Our sea levels are rising. The only difference between the global impacts and the local ones is that the latter are happening right here, on our doorstep.

So, given we're talking about the climate warming, let's get straight to the obvious: things getting hotter. In the past century, average temperatures across New Zealand have risen about 1°C, and over half of that warming has occurred since that very hot day in 1973. In the nearly 50 years since, the Canterbury record has not been topped – but some days have come very close. The summer of 2017–18 was our warmest on record, and dozens of locations from North Cape to Invercargill had their warmest-ever summers. I still remember standing on the deck at the back of my house on the Kāpiti Coast, thinking it felt like being in the tropics. It felt weird, and a bit worrying, having such warmth and humidity make it seem more like I was in Brisbane than Kāpiti. We have a wood-burner that keeps us nice and warm in the winter, but like a lot of Kiwis we don't have any air-conditioning. Hopefully we can keep it that way for a while yet.

One of the reasons our hottest-day-ever record still stands has to do with what's called climate change emergence. When we talk about something like extreme heat in a given region or territory, we have to factor in not just the rate of average warming, but also how much temperatures in that area vary naturally. The more naturally variable the temperatures, the wider that region's range, and therefore the greater the change we need to see before things move out of that natural range. In other words, how long it takes for temperatures to go off the charts depends entirely on what the charts look like. If you live somewhere like Canterbury, where temperatures tend to go both very high and very low, your temperature chart is wide. It would take a lot of average warming to push temperatures consistently beyond the top boundary of past records. If you live somewhere like the West Coast of the South Island, however, where temperatures tend to be more constant, then your narrow chart is easier to break away from. Fairly modest warming will produce record highs never observed before.

Temperatures across most of New Zealand are fairly variable, so it takes quite a lot of warming to see a trend emerge above the noise of month-to-month and year-to-year ups and downs. Generally speaking, our temperatures are most variable in the east and least variable in the west. That's due again to the country's mountainous nature, and the westerly winds. Temperatures tend to be more constant in the west with the winds coming off the slowly varying oceans. Meanwhile, in the east, the effects of downslope warming and the greater distance downstream from the oceans mean temperatures go up and down a lot more. We see a similar pattern in the rest of the world: being close to the ocean usually makes temperatures less variable, while some of the most extreme variability happens in the middle of big continents. Far from the oceans, temperatures can vary enormously, so we have to wait longer and see much more warming before the trends emerge in those places.

If temperatures weren't rising, we might expect to see about as many heat records broken as cold records – a ratio of around 1:1. However, because temperatures *are* rising, over the past ten years both here and around the globe we've seen about twice as many high-temperature records broken as low-temperature records. And, in very warm years, the ratio is much higher. During 2022, New Zealand's warmest year on record, 137 high temperature records were set in the annual statistics, while no minimum temperature records were set (apart from some in particular months), according to NIWA's annual summary.

During the 2017–18 summer, 145 high-temperature records were broken but not one low-temperature record.

As the ocean surface warms, we're also seeing marine heatwaves occur much more often than they did 50 or 100 years ago. Even the term 'marine heatwave' wasn't common in the climate science literature until the last ten years or so. During that scorching summer in 2017–18, a really strong marine heatwave made the Tasman Sea as much as 3°C warmer than normal. Tropical fish species were seen around our coasts, glacier melt went through the roof, and Marlborough grapes were ready to harvest a couple of weeks earlier than normal.

We don't hear all that much about fires in this country, but New Zealand is no stranger to them. There are an average of around 3,000 wildfires here every year, affecting around 6,000 hectares of land and costing around $100 million in damage. In recent years, there have been wildfires on the Port Hills near Christchurch, in Pigeon Valley near Nelson and, most recently, at the settlement of Ōhau in the Mackenzie Country in the South Island. This last was one of the worst wildfires recorded in New Zealand, with nearly 50 houses destroyed. Fire danger is typically relatively high in all these regions of the country, but all are seeing an increase in the fraction of the year spent in very high or extreme fire danger as the climate warms.

It comes as little surprise that, in a warming climate, New Zealand is also experiencing severe droughts. They may develop more slowly than something like a flood, but they can be every bit as damaging and extreme. The droughts that affected the country in 2007 and 2013 are estimated to have, in combination, knocked nearly $5 billion off our economy. And, using climate change attribution studies, we can say that climate change-related drought, just from those two events, has already cost the New Zealand economy around $800 million.

*

As well as seeing droughts on the rise, we're also experiencing more floods. That might seem counterintuitive, but the increase in downpours is directly related to the air temperature. Remember, warmer air can hold more water vapour, so, as the climate warms, the total amount of moisture in the air increases. This is why the tropical regions are the most humid and sticky, and is why Antarctica is officially classed as a desert – it is so cold there that there's almost no moisture in the air.

When there's a storm in a warmer climate, there's more water available to fall out of the sky, so heavier downpours come more often. In December 2011, Golden Bay flooded after Tākaka saw rainfall that was double the previous record. A decade later, in November 2020, flooding occurred in Hawke's Bay that was estimated to be a 1-in-250-year event, and left over a hundred houses in Napier uninhabitable. The winter of 2022 was both our warmest and our wettest on record, with many extreme rainfall events from June to August causing flooding and slips, and damaging roads and homes in much of the North Island and the top of the South. In August, things got so severe that a State of Emergency was declared in Nelson, Tasman, the West Coast and Marlborough, and it will likely take years to recover from the damage caused. The rainfall in Nelson, in particular, was estimated to be a 1-in-120 year event. On 23 January 2023, phenomenal rainfalls hit Auckland city with widespread flooding, slips and extensive damage. Tragically, four people lost their lives, and a state of emergency was declared. Just how unusual this storm was had not been worked out at the time of writing, but it is bound to be a very rare event based on past history, no doubt helped along by the warmer, moister, atmosphere we live in these days. The extreme rainfalls from ex-Tropical Cyclone Gabrielle were no doubt also pumped up by global warming, resulting is widespread river flooding, slips, and catastrophic damage along the east coast of the North Island, and areas farther north.

We can work out the rarity of such events by looking at past observations of weather and climate and their extremes. Of course, measuring these occurrences in terms of the average timeframe in which they're likely to take place again is something of a moveable feast. We know that, as the climate warms, rare events are becoming much more common, so what was a 1-in-120 year flood event is likely to come again much sooner in future.

*

Adding to our flood risk are rising sea levels. Just like heatwaves or floods, coastal erosion has been going on forever, but as the high-tide mark moves further inland it becomes easier for storm-driven waves to inundate coastal regions.

Along the West Coast of the South Island, roads have already been moved and public land lost to the encroaching sea. A period of strong northwesterlies in September 1976 eroded several metres of coastal land in parts of Kāpiti, causing extensive damage to properties and washing one house into the sea. That event hasn't really been matched since, but as sea levels rise extreme events become more likely, so it's only a matter of time until that 1976 event is exceeded.

Coastal engineers use the rule of thumb that 10 centimetres of sea-level rise will make any given coastal inundation event three times more likely on average. In other words, with 10 centimetres of sea-level rise, that 1-in-100-year coastal flood will instead happen once every 33 years on average. Add another 10 centimetres of rise, and it's now a 1-in-11-year event. Those numbers are just averages. In some places, the change will be faster and in others slower – it all depends on the shape and make-up of the beaches and coastal margins.

When it comes to talking about sea-level rise, it's important to know whether the land you're on is rising or sinking, as this can either cancel out the problem or make it worse. We also have to factor in the world's big ice sheets –

since they have their own gravitational pull, they attract ocean water and pull up sea levels around their coasts. As ice melts, the local sea level actually drops because of the reduced gravity. The meltwater flows far away and the net result is that melting in Greenland raises sea levels more in the Southern Hemisphere, and melting in Antarctica raises sea levels more in the Northern Hemisphere. As the earth's crust adjusts to the changing weight of water, sea levels can change differently in different regions.

New Zealand is a dynamic land, sitting on the boundary between the Pacific and Australian tectonic plates, so things vary from region to region. Adding to our coastal woes, we're experiencing similar land-subsidence issues to cities such as Shanghai or Jakarta, but here it's not so much a result of urban development as of the geologically active nature of our country. Recent work by NZ SeaRise: Te Tai Pari O Aotearoa has shown that many parts of New Zealand are experiencing subsidence at a rate of up to a few millimetres a year. That effectively doubles the local rate of sea-level rise in some locations, such as Wellington, bringing increased coastal flooding and erosion much faster than we thought might occur only a few years ago.

All told, around Wellington the land is sinking at about 3 millimetres per year, roughly doubling the local rate of sea-level rise. In parts of the Bay of Plenty, on the other hand, the land is rising about 3 millimetres per year, roughly cancelling out local sea-level rise. And, generally speaking, over the past century, New Zealand sea levels have already risen around 20 centimetres.

For Wellington, the increase in the occurrence of coastal inundation events has been estimated as a factor of five for every 10 centimetres of sea-level rise. That means that, with 50 centimetres of sea-level rise around Wellington, a 1-in-100-year coastal flood would be occurring every couple of weeks. With 80 centimetres of sea-level rise, it would be happening on every tide.

*

Winds play a pivotal role in New Zealand's climate, and just like everything else they're also changing – but how, exactly, is a complex story.

Winds blow along the isobars between areas of high and low air pressure, the intensity of the highs and lows, and the strength of the winds in between, depends on differences in temperature. The bigger the temperature difference, the stronger the winds. So it's not just overall global warming that's at play – the geographical pattern of regional warming also has to be taken into account. Like many aspects of climate change, what's happening with the winds really depends where you look.

Things are also further complicated here in New Zealand by our most important pattern of climate variability, the SAM. Ordinarily, as a result of the very slow warming of Antarctica and the tendency for the Southern Oceans to mix heat down to well below the surface, we'd expect to see stronger westerly winds on average. The change in wind speeds would be relatively small – just a few per cent over several decades – but that would still translate to an increase in rainfall of a few per cent in the west and a similar decrease in the east, because of the effect of our mountains. However, because of the SAM and its relationship with the ozone hole, the westerlies over New Zealand have actually been *decreasing* in recent decades. The trend for more positive phases of the SAM has offset any increase that would be expected from the global-warming pattern. In addition, the trend has actually also helped warm New Zealand, as the positive SAM goes with warm conditions over this country. And, at the same time, winds over the Southern Oceans have become stronger, and that has caused the Antarctic Circumpolar Current – the one current that flows all the way around the globe – to move faster.

As the ozone hole recovers, it's likely the positive SAM trend will disappear, or even reverse if we can reduce greenhouse gas emissions fast enough. That would, ultimately, mean New Zealand would experience a few more weeks of windier and unsettled weather each year than we've become used to in the last three decades – but I'd say that's a small price to pay if it means we're getting climate change under control.

*

When I was eight years old, my aunt told me a tall tale about life back in the days when she lived on a high-country sheep station. Each night, she would head down to the farm gate to leave an enamel jug for the travelling milkman to fill overnight. One morning, she said, the frost was so hard that all she found at the gate was a jug-shaped block of frozen milk. Nearby lay the shattered remnants of the enamel jug. 'It had been split in two by the freezing milk expanding!' she told me, and I absolutely believed her (although, later in life, I've had a few questions).

Cold and frosts coloured the winters of my childhood. Around the same time as my aunt told me her milk-jug story, central Canterbury experienced a storm that dumped half a metre or so of snow across the region. My primary school was closed, as were several others, but my dad had a caretaking job at the school so he took me along with him, and I got to spend the day with the snowy grounds all to myself. The snow stuck around for days, and when school finally reopened the snowball fights and snowmen were epic!

In the winter of 1939, a cold outbreak brought snow to all parts of the country – and I mean all parts. The white stuff covered the length of the country. Snow settled on Mount Eden in Auckland, and on the hills behind Kaikohe in the Far North. It even fell at North Cape. The snowfalls were heaviest in the south, lying up to a metre deep in suburbs of Dunedin.

Even more incredible, back in 1895, the Waitaki River froze over. The Waitaki is no creek – it's one of the South Island's major braided rivers, responsible for feeding water into several of the key dams in the country's hydroelectricity system. Looking at it today, it is hard to imagine the whole thing freezing over, and yet that's exactly what happened. There is even a photograph in the Kurow museum of the local police constable standing on the ice to prove it. Imagine how cold it would have had to be for something like that to happen! Nothing has come close in my lifetime – but, going by how casually that policeman was standing on the ice, it may not have been such a rare event in 1895.

Snow, ice and frosts are all becoming rarer these days. I cannot count the number of times someone has come up to me after I've given a public talk to remark on the warming they've noticed in their own lifetimes. 'When I was a kid, we all smashed the ice in puddles on the way to school,' they tell me. 'Now, there's hardly ever ice in the puddles, and my kids don't have that experience.' And in this case their memories are serving them well: the occurrence of frosts has gone down dramatically in the past 50 years. Most places now see fewer than half as many frost days, and the explanation is pretty simple: just as warming the whole climate makes it easier to get hot days, it also becomes harder to have cold nights.

To get a frost, the air temperature needs to fall below 0°C, but it also helps to have still air and clear skies, and preferably low humidity. At night, the air is cooled from below (just as it is warmed from below during the day). This is because, when the sun disappears in the evening and the solar warmth switches off, the remaining heat energy in the air isn't enough to balance the loss of heat from the ground. And, the clearer and drier the air, the more easily heat escapes out to space, and the faster the ground cools. So, for a few metres above the ground, temperatures can actually increase with height – a situation called an inversion, because it inverts the usual way of things, where the warmest air is at ground level. It's a very stable situation, where the air resists rising and falling, and the cold layer can become isolated from the atmosphere above. This forms what's known as a cold boundary layer.

If there's any wind around, however, an inversion and a cold boundary layer cannot form, because the warmer air above keeps getting mixed with the cold air near the ground. That's why grape growers and other horticulturalists sometimes use helicopters on cold nights to mix up the warm and cold air and stop frost from damaging their crops. It works, but it is very expensive!

On a cold night with an inversion, the coldest place is the ground itself. If the ground temperature falls below 0°C, that's what is known as a ground frost. If the air temperature a metre or so above the ground also falls below zero, you'll get an air frost. On cold and still nights, the difference in temperature between the ground and the air just a metre or two above can be as much as 4°C, meaning ground frosts are much more common than air frosts in most places in New Zealand.

When most scientists use the word 'frost', they're talking about an air frost, and that's when you see the most damage.

It's not just grapes and apple blossoms that suffer in a frost, either. Many cold winter nights when I was growing up, I remember Dad turning off the water mains and running the taps in the kitchen until they went dry. He did this because, if the water froze in the pipes, they might burst. They never did, but some mornings when we went to turn the tap on so we could brush our teeth, only a frustrating trickle would emerge. Plugs of ice had formed in the pipes and the water couldn't flow past them. We had to wait until the ice melted – usually around lunchtime – before the taps worked properly again.

Nowadays, waiting for ice to melt is not one of our problems. As the climate warms, winters become shorter, more precipitation falls as rain rather than snow, and our cold bits – our glaciers – are melting. Since the 1970s, the glaciers in the Southern Alps have lost about one-third of their ice. In the early nineties, a lake formed at the terminus of the Tasman Glacier and has been eating away at the ice ever since. At the same time the Tasman, New Zealand's biggest glacier, has been thinning substantially. The Fox and the Franz Josef are going the same way, as are all New Zealand glaciers.

We know how much ice has been lost from New Zealand glaciers over the past 40 years because of the pioneering work of glaciologist Trevor Chinn. In the seventies, he hit upon the idea of photographing the Southern Alps glaciers from an aircraft in order to judge changes in the amount of ice held in them. He theorised that, if the images were taken at the end of summer, when the seasonal snow-pack was at its minimum, it would be possible to see how the permanent snowline was changing and from there work out changes in ice volume. The flights got under way in 1977, and the glacier images taken every year since then have provided a unique record of change in parts of the country that are particularly inaccessible. Trevor died in 2018, but the survey flights carry on, supported by NIWA and Victoria University of Wellington.

As for snow, we might expect that big dumps would become a thing of the past in a warming climate, but snowfall is still common enough in the south and east of the South Island. A dusting usually shows up once or twice during most winters, with bigger falls every few years. Much of the South Island is cold enough in winter that snow is still possible in a warming climate, for a while at least. Even though the atmosphere and the oceans are warmer now than they were a century ago, temperatures at ground level can still be below freezing in cold southerly outbreaks. With the warmer air coming across the Tasman from the northwest, snowfalls these days can actually be heavier than they were in the past, because warmer air contains more moisture.

In the North Island, however, snow to low levels is much rarer and temperatures are often more marginal – things are cold, but often above freezing. The winter of 2022 and the couple before that were not very snowy at all for the North Island. That's partly the result of the warming climate, but also the effect of the long-lived La Niña event in the tropical Pacific. A La Niña winter usually goes with warmer air, more northerly winds and less snow generally, and that's pretty much what we've seen.

Ski operators have known for many years what they're up against with climate change. Over time, the economics and logistics of skiing are becoming increasingly marginal. In lieu of natural snowfall, ski fields can use snow-making machines to extend the length of the season and keep the slopes covered, but even then it still has to be cold enough – ideally below -2°C, with low humidity – and those conditions are becoming rarer over time. The 2022 winter put this vulnerability on show when Ruapehu Alpine Lifts, the owner of Whakapapa and Tūroa ski fields, went into voluntary administration. We can't pin it all on the warming climate – the Covid-19 lockdowns and back-to-back La Niña events also played their part in the ski fields' difficulties – but one thing is certain. As time goes on, climate change will be a more and more important factor.

7

Our big blue backyard

The peopling of the Pacific Islands is an epic story. Small bands of sailors set off across vast stretches of ocean to find the next island group, reading their way by the stars, the sea surface, the wave patterns and the winds. These mariners – among the world's very best – were capable of expertly tracking a course from one tiny atoll to the next small volcanic pimple, thousands of kilometres away. They found their way to Aotearoa's shores long before anyone else. Forget sextants, compasses, GPS. No need for those when you have tools like stick charts to encapsulate swell and wave patterns across an ocean region, charts that map the pathways between islands and speak of a deep knowledge of ocean behaviour.

Over the centuries, Pacific people have perfected the art and skill of navigating our big blue backyard. The ocean has always played a leading role in these communities: it's a major source of food, a conduit for travel, and a presence that gives meaning to life in the islands. But climate change is affecting these oceans, and that's posing huge risks for all Pacific Island nations.

I spend a lot of time talking to students, politicians and journalists about the impacts of climate change here in Aotearoa New Zealand, but in fact our Pacific neighbours are the ones who are truly at the forefront of climate change. Across the western Pacific, sea levels are rising faster than the global average. Temperature extremes are going off the charts faster than in many other places, rainfalls are increasingly variable, and tropical cyclones are becoming more intense. These nations are right in the eye of the storm.

Over 10 million people call the Pacific Islands home. Here, ocean surface is a better definition of a country's space than land is. Many of the region's countries are not one island, but a collection of many tiny ones, flung across the water. For instance, Kiribati's islands comprise a total land area of just over 800 square kilometres, but are spread over 3.5 million square kilometres of ocean. Many of these islands are also low lying, and therefore vulnerable to even modest rises in sea levels. In Fiji, with a population of nearly a million, over 40 coastal villages have already been earmarked for evacuation due to saltwater inundation that regularly floods houses and destroys crops and freshwater sources.

Homes, food production, freshwater, roads and other infrastructure – everything is under threat, and in many places livelihoods are already being washed away by the waves. We hear the terms 'climate mobility' or 'portable sovereignty' used to describe the need for communities such as these to relocate as the waters rise, but the euphemism hides the true gravity of the situation. What does it mean when a whole nation has to pick up and leave the land they've called home for as long as anyone can remember? What about when not one nation but many must relocate? And how do you reckon with your country being swallowed up by the very same force – the ocean – that has been your way of life for so many centuries?

Sure, it is possible to build sea walls. It might even be possible, in some cases, to move inland to higher ground. But these are usually short-term measures. Some communities and countries are already exploring what it would mean to sail away from their homes and never return.

*

When we think of sea-level rise and Pacific Island nations, many of us imagine whole islands – and even whole countries – going under the waves, but in fact islands don't just gradually submerge as the sea level rises.

Currently, sea levels are rising inexorably across the Pacific – just as they are across the world – but the change is relatively slow, at a few millimetres per year. The pattern of how sea levels are rising is related to how the increasing weight of water in the oceans pushes on the Earth's crust, and the changing gravitational pull of the big ice sheets. The net effect for the Pacific Basin is a faster rate of sea-level rise in the Western Pacific and a slower rate in the east, nearer to South America.

So far, there have been some winners and some losers across the Pacific Islands. While one or two small islands have already disappeared, others have actually grown. This has to do with the way surface ocean currents transport sediment, removing it from some islands and depositing it on others. As sea levels rise, erosion rates and ocean currents change, shifting the regional pattern of sediment movement so that some islands are enlarged, while others are eaten away. To get a clear picture for each island in each nation, we need to look at the shape of the sea floor, the role of coral reefs, and how ocean currents affect each location.

Further complicating things in the tropical Pacific are the trade winds, which blow around the global tropics from east to west. As these winds blow, they drag surface water with them, and pile it up in the west of the ocean basins. Across an ocean as wide as the Pacific, that means an average difference in sea level between South America and Indonesia of as much as 20 centimetres. Changes from year to year in the strength of the trade winds (related to the El Niño–La Niña cycle) can add to the background rising trend, or cancel it out, or even reverse it, for years or even decades at a time.

For now at least, while the rate of sea-level rise is a matter of millimetres, Pacific nations have some time to plan and adapt – and grapple with other changes under the waves, notably the rising temperatures and the acidification of ocean waters. Marine heatwaves are on the rise in the western Pacific, as they are in all of the world's oceans. Ocean heat and increasing acidity put tropical coral under a lot of stress. The expectation is that tropical coral reefs will all die out at 2°C of global warming, though a fraction would likely be saved if we can stop the warming at 1.5°C. Fish species are very sensitive to temperature change, meaning that key commercial species could disappear, or at least become a lot rarer across the Pacific.

Earlier, we talked about climate change emergence – the idea that the amount of time it takes for temperatures to go off the charts depends entirely on what the charts look like. Well, unlike Canterbury, the tropical regions don't tend to have very wide charts. The sun shines strongly in the tropics all year round, and it is always warm, meaning temperatures barely vary. Residents of Apia in Samoa, for instance, can depend on day-time temperatures being within a degree of 30°C every month, every year. If the temperature reaches 32 or 33°C, that's remarkable.

That means the warmest places on Earth are also the places where the warming signal is emerging the fastest. In many parts of the tropics, high-temperature extremes are likely to go beyond anything observed for centuries within the next two decades or so. And, while some residents might be able to just crank up the air-conditioning, that's not the case for most living in many tropical countries. Tropical ecosystems like coral reefs certainly don't have that luxury.

*

We sometimes hear that there are an increasing number of tropical cyclones and they are becoming more intense, but that's not exactly right. Actually, the total number of tropical cyclones (also known as typhoons or hurricanes) is gradually decreasing worldwide – a fact not widely understood outside the climate science community, but clearly demonstrated by all the models and observations. However, at the same time, the fraction of very intense tropical cyclones is on the rise. So, in other words, there are fewer storms overall, but when they do come they are more devastating, with stronger winds and more intense rainfalls.

Tropical cyclones get their energy directly from the ocean's surface, so they need sea surface temperatures of at least 26.5°C to form and develop. The cyclone season in the South Pacific normally runs from November to April, the warmest months of the year, but a few have been known to form outside of this period. These storms are very different from the ones that routinely cross New Zealand, which tend to have the coldest air near the centre and are fuelled by the difference in temperature between the warm and cold air masses. A tropical cyclone, by contrast, is warmest in its core and gets gradually cooler outwards from there. The most intense storms have a clear patch in the very middle: the eye of the storm. Generally speaking, the bigger the eye, the more devastating the cyclone. Some weaker tropical cyclones don't develop a clear eye, and instead clouds extend over the whole storm.

Like all storms, the winds in a tropical cyclone rotate around the centre. As air is drawn towards the centre of the developing storm, it feels the rotation of the Earth beneath and starts to take a curved path. The result is that winds blow clockwise around the centre of a storm in the Southern Hemisphere, and anti-clockwise in the Northern Hemisphere. However, that rotation-related Coriolis effect is not felt right on the Equator, so you have to be about 5 degrees latitude north or south before the Coriolis effect is strong enough to spin up a tropical cyclone.

As the oceans warm, the region where surface waters are above the 26.5°C threshold is expanding polewards. It is already possible to measure that tropical cyclones are developing farther from the Equator than they used to a few decades ago. The speed of movement of tropical cyclones appears to be decreasing gradually, because cyclones are steered by the winds blowing across the tropics and wind strengths in the tropics are gradually decreasing. This is not good news for islands in the path of tropical cyclones, as more intense storms that move more slowly are likely to do more damage.

One of the defining features of the southwest Pacific climate – and a strong controller of tropical cyclone activity – is the South Pacific Convergence Zone (SPCZ). While it was first identified in early satellite photographs during the 1960s, it has no doubt been familiar to Pacific peoples for centuries. This band of intense cloudiness and rainfall lies from the region of Papua New Guinea southeast towards French Polynesia, petering out and merging with mid-latitude weather systems out over the central South Pacific. Since the heaviest rains fall beneath the SPCZ across the southwest Pacific, it's an important source of freshwater for a number of Pacific nations.

Despite the fact it's been studied for 50 years or more, the SPCZ remains a little mysterious. It really only exists during the wet season, and from one month to the next – even from one year to the next – it can lie across a wide range of latitudes within the southwest Pacific, the main control on its movement being the El Niño/La Niña cycle. At its western end (nearest to Papua New Guinea and Australia) the SPCZ doesn't move north and south all that much, but farther east its tail wags quite dramatically across the central Pacific. Where it lies exactly in a given year is really important for regional climates. If it moves south and west, island groups farther north, such as Samoa and the Cook Islands, can experience drought, while Fiji, Tonga and Vanuatu can be affected by torrential rain and flooding. The opposite occurs if the SPCZ drifts to the northeast.

In the South Pacific, tropical cyclones only ever form south of the SPCZ – either near the SPCZ itself or within a few degrees of latitude to the south – and move more freely in El Niño years, when the SPCZ tends to be farther north and east of its normal position. In really strong El Niño summers, such as those that came at the end of 1982 and 1997, the SPCZ can swing completely out of the way and lie almost completely east–west, opening up the whole of the tropical southwest Pacific to cyclone danger.

The location of the SPCZ also dictates whether floods or drought will be the 'theme of the year' for a number of South Pacific island nations. In El Niño years, when the SPCZ moves northeast, countries in the southwest like Fiji, Niue and Tonga can experience severe drought, while Samoa and Tuvalu may be on the receiving end of damaging floods. The situation reverses during La Niña. And, as the climate warms, we're expected to see more 'super El Niños', while the swings of the SPCZ associated with the El Niño cycle are expected to get larger. Whatever happens with El Niño, it's clear that rainfalls are becoming more variable across the Pacific, putting more stress on freshwater supplies as time goes on.

*

Fiji, a beloved holiday destination for Kiwis, is just one of the Pacific nations already suffering the effects of climate change. Since 2016, the country has been hit by 13 natural disasters, ranging from cyclones and floods to intense droughts. There are other problems too, such as ocean acidification, which is damaging coral reefs and putting stress on the entire marine ecosystem.

While Fiji is no stranger to cyclones, the storms are growing in intensity and hammering the country on all fronts: its infrastructure, its housing, its farms, its fisheries and its culture. In February 2016, category five super cyclone Winston made landfall in Fiji as the Southern Hemisphere's strongest-ever recorded storm. It knocked out power to 80 per cent of the population, destroyed nearly 800 homes and damaged 40,000, killing 44 people in the process.

The town of Vunidogoloa on Vanua Levu, the second-largest island of Fiji, provides a potent example of what is already happening to many Pacific islands. Back in 2006, the local residents started expressing alarm about the sea washing through their villages – they regularly found themselves knee-deep in seawater, and salt was leaching into the soil, ruining their crop beds. However, it took until 2014 before they were officially relocated to new homes 1.5 kilometres inland and out of harm's way. Since they were no longer able to fish directly in the sea, ponds were created for them to farm fish in, and the cemetery was also shifted to be near the new village, as the ancestral burial grounds succumbed to the sea.

While the relocation was considered by Fiji's officials to be a success in practical terms, the trauma that residents felt at having to leave their land lingers. Patricia Mallam, an environmental activist turned climate-adaptation researcher based in Suva, told me that the sense of dislocation those relocated from Vunidogoloa and other villages feel has had profound emotional consequences. 'Relocation should be the last resort,' she explains. 'If that's something we are arriving at now, the future does look kind of bleak.'

*

It's a particularly cruel aspect of climate change that many of those being most affected by it are doing the least of the damage. Our Pacific neighbours are just one such example of the inequalities of climate change.

It might, therefore, be tempting to cast Pacific nations as victims of climate change, as vulnerable, but to use these labels without further context does a disservice to these communities that have existed for centuries – and it also gives countries like New Zealand a convenient excuse for looking the other way. The reality is this: despite being some of the smallest contributors to climate change, and despite carrying very little economic weight on the global stage, Pacific Island nations have worked harder than most at international negotiating tables. The Pacific island states, the Pacific Islands Forum and the Alliance of Small Island States (AOSIS) have all been at the forefront of action on climate change for many years, at policy and political levels. In the words of the Pacific Climate Warriors, a grassroots movement that's actively demanded climate justice for two decades: 'We are not drowning. We are fighting.'

Fiji and other Pacific island countries are already in adaptation mode, building sea walls and moving infrastructure out of the way of storm surges. They are also cutting their own emissions through building renewable-energy plants to replace fossil fuels. In 2019, the Fiji Government created the Climate Relocation and Displaced Peoples Trust Fund for Communities and Infrastructure, the first fund of its kind in the world. Its core funding amounts to around FJ$5 million a year, taken from the Environment and Climate Adaptation Levy, which is collected from Fijian businesses. In 2020 New Zealand pledged $2 million to the fund, part of a $150 million package to pay for climate resilience projects across the Pacific region. 'We need to arm ourselves with the ability to act now,' Fiji's Prime Minister Frank Bainimarama said when the relocation fund was launched. 'We can't wait for communities to be drowned out by the encroaching tides. We need a holistic approach, we need adequate resources and we need it now.'

But there's only so much that these poorly resourced nations can do on their own. Real action requires participation from the international community, and that includes us here in New Zealand. While the United Nations helps fund climate-adaptation efforts around the Pacific, there's currently precious little funding going into relocation of climate refugees. Meanwhile, a 2016 study commissioned by the New Zealand Ministry of Foreign Affairs and Trade estimated that 'no atoll group in the Pacific is likely to be habitable by the end of the century'. Island nations like Nauru, Tuvalu and Kiribati face the most immediate threats, but the impacts of erosion, storm surges, a lack of freshwater and drought could force the inhabitants to leave long before their islands are close to being submerged. So major questions hang in the air over the legal status and sovereignty of people who are forced to leave their homeland entirely to escape the impacts of climate change – impacts that they, as nations, have done very little to cause.

Back in 1990, the AOSIS was established to give small island and low-lying coastal developing states from around the world an international voice on climate change and sustainable development negotiations. Of its 39 members, 15 come from the Pacific. The alliance played a large role in the design of the UN Framework Convention on Climate Change, and has had a strong voice throughout the COP meetings held annually since the mid-1990s. In particular, AOSIS very effectively lobbied the international community to set more ambitious targets for emissions reductions, including pushing for the world community to aim for no more than 1.5°C of global warming, instead of the proposed 2°C limit.

At the same time, it has been Pacific countries pushing most strongly for loss-and-damage funding, something first proposed by Vanuatu in the early 1990s. This was one of the main topics of conversation at COP27 in Egypt, the 2022 meeting of the parties to the UNFCCC: how will 'loss and damage' be paid for, and who will provide the funding? By the end of the COP, a mechanism had been set up to allow funds to flow from developed countries that have caused the problem to developing countries dealing with the effects. The exact details of who will pay, and how much, have yet to be worked out – and are very contentious issues, sadly.

This is part of a broader conversation about climate justice, whereby the fact that the countries that have done the least to cause the problem of climate change are bearing the brunt of the consequences, and this only increases the inequities between countries. The hope is that taking action on climate change can be done in such a way as to reduce inequity, rather than to increase it. That makes a lot of sense to me and does indeed feel like justice.

The loss-and-damages agreement reached at the end of COP27 is a real win for the Pacific, after three decades of tireless advocacy. However, the goal of limiting warming to 1.5°C was not pursued with much vigour at the latest COP – but, yet again, it was Pacific Island nations pushing hardest for action. At the meeting in Sharm el-Sheikh in Egypt, Tuvalu put on the table a suggestion that the countries of the world should set up a global treaty to phase out the use of fossil fuels. That kind of action is just what is needed to wind down the emissions of carbon dioxide. The science tells us that, unless emissions are cut drastically through the 2020s, global warming will exceed 1.5°C and many Pacific Island countries will be at much greater risk from sea-level rise.

Here in New Zealand, we have much to learn from our neighbours in the Pacific. They don't need our sympathy or our pity. They need us to join with them to take up the fight. Ultimately, the best way any developed country can demonstrate support for places like the Pacific Islands is to cut greenhouse gas emissions as fast as possible. If all the nations on the Pacific rim – New Zealand, Australia, China, Japan, Canada and the USA – took urgent action on emissions reductions, then the Pacific really would have cause to celebrate.

III

THE FORECAST

8

Good, bad or ugly?

I'll never forget the first time I heard David Bowie's 'Moonage Daydream'. I was 15, it was the school holidays, and I was listening on Radio 3ZM. It sounded pretty far out and other-worldly to this country boy from central Canterbury. I was hooked.

I went to buy the LP, *The Rise and Fall of Ziggy Stardust and the Spiders from Mars*, at the Record Factory on Colombo Street in Christchurch. When I emerged from the store, I held it before me like a sacred relic, and at home I pored over the image on the cover: Bowie beneath a violet night sky, standing outside a dingy brick building with one purple-booted foot hitched up on a cardboard box before him. A streetlight glows above him, while the pavement beneath glistens with rain. There's a sign above his head reading 'K. West' – what did that mean? I wondered. I played that album on our little radiogram over and over, all summer long. I'm not sure what Dad thought of Ziggy Stardust, but he remained polite – and, in return, played the soundtracks from his beloved Broadway musicals, *South Pacific* and *My Fair Lady,* just as much.

I was so taken with the album's story of a rock star from outer space, sent to Earth to avert an apocalypse. In the opening track, 'Five Years', Bowie sang about how the 'Earth was really dying' and we had 'five years left to cry in'. Back in the 1970s, that just seemed cool, space-agey sci-fi, but I didn't think it had any connection to the real world. Listening to it today, I'm not so sure. There's no apocalypse coming from outer space (hopefully) and the end of everything in five years is still not on the cards (also hopefully), but we have managed to cook up a home-grown crisis that could lead to a great deal of crying if we don't take decisive action.

So, how bad could it get? How fast can things change?

Take a look online and you'll find every possible opinion about the future, if you look hard enough. At one end of the scale, you'll be told all of humanity will be extinct by 2030 or sooner, while at the other end you'll hear climate change is actually a wonderful thing and we all have a happy and healthy future ahead of us as far as the eye can see. The truth, as it has a tendency to do, most likely lies somewhere between those two extremes.

The thing is that there is no magical temperature at which catastrophe happens. There's no specific point at which we'll all die, or climate change gets into a runaway state. We're the ones who control how much greenhouse gas we're emitting, and that's what dictates the rate of climate change – so we have the power to stop it (or not). It can help to think of the climate system as a train on the tracks. The whole of humanity is in the carriages, with some of us making the train go by shovelling coal into the furnace. The more coal we shovel, the faster the train goes, and the bigger the speed wobbles. At some point, the carriages are going to start to derail. What do we do at that point? Do we keep shovelling coal? Or do we stop?

Working out what's coming requires us to know two things: what we'll do with greenhouse gas emissions, and what changes are in store. Climate models are a vital tool here, but they are only part of the story. We can also look at past climate changes to see what might be possible, though none of it can really give us a perfect analogue for the future. One of the main reasons for that is the pace of change. There have been rapid changes in climate in the past – along with mass extinctions, and huge rises and falls in sea level – but nothing we know of that's caused carbon dioxide levels to rise as fast as they are now. We don't know exactly how quickly the oceans and the ice sheets can respond. That's something we may just have to find out as we go.

Some like to use the fact the climate has changed in the past as a reason not to worry about what's happening today. Personally, I don't get a lot of comfort from knowing that in the past the world has gone from frozen to tropical and back again, that sea levels have risen and fallen a hundred metres, that mega-droughts and deluges have come and gone. And one important thing that all of these past climates have shown us is that the climate system is sensitive. Very sensitive. And the two things it is *really* sensitive to? The brightness of the sun and the amount of greenhouse gases in the air. So when I think about exactly how much the climate's changed in the past and the fact we've pushed carbon dioxide levels beyond anything known for three million years, I do not feel a cause for complacency!

Humanity was lucky to come through the ice age cycles and emerge into a global climate perfect for sustaining life, one that has been remarkably stable for several thousand years. That climate allowed us to invent agriculture, cities, money, smartphones, YouTube. It gave us the luxury of time, allowing us to build this wonderfully complex technological world we live in – but the downside is that all the fossil fuels we've burned doing so have brought that lovely period of climate stability to an end.

Don't get me wrong: the climate will stabilise again, once we stop this experiment we're doing with our life-support system, but that new climate will be very different to what we've known. If we want to stop at a point that is manageable for most of us, the luxury of time is something we no longer have.

*

The best way we have to judge what the future holds is to use climate models. First, we need to simulate a future with no greenhouse gas increase – these are the natural simulations. Then, we simulate the future with a range of greenhouse gas levels – the anthropogenic (human-caused) simulations. Then, finally, we compare the natural simulations with the anthropogenic ones.

To run forward into the future, models need to know how much carbon dioxide and other greenhouse gases will be in the air every year from now until the end of the model run (maybe 2100, or 2200, or even 2300). Since we don't know exactly how fossil-fuel use and renewable technology will evolve in future, climate modellers run a range of 'what-if' futures to capture the range of what's possible, or what's likely.

A 'green' future might see greenhouse gas emissions come down to zero quickly, with very little further warming. We can even look at what would happen if emissions stopped overnight (wouldn't that be amazing!). At the other end of the possible futures, we might imagine digging up all the coal and oil we can find and burning the lot.

There are a handful of standard scenarios – in between stopping emissions right away and burning everything – that climate modellers all over the world use to look at what the future might be like with different amounts of greenhouse gases in the air. By comparing many different models, and even by comparing different simulations with the same model, we can average out a lot of the noise in the modelled climate – everything from the daily weather to the comings and goings of El Niño and La Niña. By doing so, we get a much clearer picture of the warming signal driven by greenhouse gases.

One of the most striking things climate models demonstrate is that, whatever the future for greenhouse gases, the pattern of change is very consistent across different models. We see some of what we are already noticing in the observations – the things already outlined in previous chapters. There's the greatest warming in the Arctic, the least around the Antarctic; there's drying in the desert regions that are already dry, and wetting in many other places, especially the already wet monsoon regions in the tropics.

The reality is that, barring miraculous reductions in emissions in the 2020s, the globe is committed to at least 1.5°C of warming. What matters is the overall amount of change beyond that. That's what tells us whether the future will be good, bad or ugly.

9

New Zealand's outlook

In the 1980s, when the threat of nuclear war became a serious worry, many US billionaires picked out New Zealand as a safe haven. They seemed to think it would be a good place to own property if things went bad at home. Wherever the bombs were dropped, the theory went, New Zealand would be far from the action, so would be protected from the fallout and the after-effects.

These days, there are still those who think New Zealand is a safe haven, but the threat they're fleeing has become climate change. Down here in the South Pacific, we're surrounded by oceans and far from big land masses. We enjoy a Goldilocks climate that's not too hot and not too cold. But could New Zealand really become a climate change oasis in the future?

To a certain extent, we do have some advantages in a changing climate. Since the westerly winds and the competing effects from the tropics and the South Pole are not going away, our local climate is likely to stay fairly spring-like, windy and well watered in most areas. The nature of the day-to-day weather – the mix of highs and lows, sunshine and rain – is not likely to change noticeably over the rest of this century.

Since our present climate is temperate, it has to change quite a bit before things become properly problematic here. Compare that with Australia, the land of flood and fire, where extremes of weather and climate are routine fare and have been for thousands of years. Southeastern Australia, in particular, is in line to see even more extreme heat this century. Already, Melbourne and Sydney experience temperatures over 40°C, and research shows that, with only another degree or so of average warming, extremes above 50°C are likely well before the end of the century. As well as causing major public-health issues, those sorts of temperatures in the major cities of our closest neighbour will likely put strain on electricity supplies and transportation systems, as millions of people crank up their air-conditioning, train lines buckle and roads melt. It's not out of the question to imagine Sydneysiders and Melburnians choosing to move away to escape the heat, and New Zealand could well fall in their sights.

What happens beyond the next couple of decades depends on what the world as a whole decides to do. If the global community does not take urgent action to reduce emissions, the level of large-scale warming and climate change will push even New Zealand's climate into a state that would make many aspects of our current way of life unsustainable. Even in a climatically lucky country like ours, 3°C of global warming would be catastrophic.

*

We already know that climate change is leading to faster warming to the north of New Zealand, so the average strength of the westerly winds is gradually increasing, and will continue to do so through at least the rest of this century. It doesn't look as though the average direction of the westerlies will change in future, so the warming that's expected is coming purely from increased greenhouse gases. The pattern of warming over the country is more rapid in the north and a little slower in the south, where the Southern

Oceans are keeping things cooler. The overall rate of warming is expected to be a bit less than the global rate, because of the slower warming of the Southern Oceans.

Stronger westerlies paint a general picture of a wet west and a dry east, with drying also in the Far North. The western parts of the country, especially the western South Island and across Southland, will experience gradually increasing average rainfalls. Meanwhile, in the eastern regions, especially those in the North Island, rainfalls are expected to decrease on average. The area from Auckland north is expected to become gradually drier over the next century, as the main subtropical high-pressure belt inches towards the South Pole. Northland will likely see more sunny skies and less rainfall in future, which means higher chances of drought.

There are a few things going on to make the drier parts of New Zealand even drier. As the subtropical high expands polewards, it will lie more often over Northland – the part of the country closest to the tropics – bringing clear skies and dry weather. Meanwhile, the eastern parts of the country, from Gisborne to Otago, are already relatively dry because they sit in the rain-shadow of the mountain ranges running the length of the country. As our westerly winds gradually increase, that rain-shadow effect will happen more often, leading to more dry and warm days in the east. With just a couple of degrees of average warming, places like Canterbury, Hawke's Bay and Northland will see a 5 to 10 per cent drop in average rainfall. And that, combined with a warmer climate, could mean a doubling or tripling of the occurrence of droughts in those regions.

In Canterbury in recent decades, demand for water has skyrocketed with the boom in dairy farming. But, while water may have been abundant enough in the region half a century ago, that is no longer the case. All that irrigation is depleting groundwater and drying up rivers, and as average conditions get drier the problem's only going to worsen, because pastures will need even more water. In some parts of Canterbury, water is already over-allocated because historic permits were handed out willy-nilly on the basis that there was plenty of water to go around. There's not. And even if today's water allocations could be sustainable, they won't be in a drier future. It seems obvious to me that Canterbury will, at some point, have to revert to a more dryland-style agriculture – the sort I knew when I was growing up. Wheat and some sheep, but hardly a dairy cow in sight.

The same goes for the other eastern regions of both islands. As water stresses mount, water use will have to be reduced, as it's not just animals and crops that use water. We do too, for drinking and for living our lives. Our increasing population will only add more pressure.

In some places, water-storage facilities are being built, and that may help, but the most sustainable approach is to work with the local climate: accept that the east is dry and don't try to irrigate it into lush pastures everywhere.

*

As parts of the country see more frequent and intense droughts, fire danger is also increasing, especially during the warmer months of the year. So does that mean New Zealand should get ready to become a land of raging fires? The short answer is no.

While fire danger here is on the rise, our climate is more temperate than places like Australia or southern California, and we have larger and more reliable rainfalls. Our location midway between the Equator and the South Pole, along with the fact we're a group of smallish islands surrounded by vast oceans, means our climate will never become as extreme as other regions.

That doesn't mean we get off scot-free, though. Some parts of the country are naturally very dry and warm, and these areas will see a greatly increased fire danger in a warmer world. With a couple of degrees of warming, many places – from Dunedin to Picton in the south, and from Martinborough to Gisborne in the north – are expected to see a doubling or tripling of the period spent every year in very high or extreme fire danger. That doesn't mean there will definitely be more fires, but it does mean the potential is there.

It all scales with the level of warming globally. If the globe warms only another half a degree from where we are now in the early 2020s, fire danger is not expected to increase so much. But, if global warming reaches 3°C or more, the period of extreme fire danger in our eastern regions may be four or five times longer than in the late twentieth century. Some areas will be in very high or extreme fire danger virtually all year round. And, worldwide, fire occurrence will likely at least double from where it was 30 or 40 years ago.

New Zealand's climate might be less fire-friendly than Australia's, but fires here can be just as devastating and hard to contain as across the ditch and elsewhere. In 2017, the Port Hills fire near Christchurch took over two months to put out. Around 1,400 people were evacuated from their homes. Nine homes ended up destroyed, and five more were damaged. Insurance claims totalled around $18 million. At the time, it was the worst fire in New Zealand for a century, in terms of damage to private homes. Then, in October 2020, came the Lake Ōhau fire, which burned through very dry country adjacent to the lake and occurred during very strong winds, so it swept quickly through the village, destroying 48 structures and burning more than 5,000 hectares of land, damaging power lines, fencing and other infrastructure. It took nine days to put out completely. Thankfully, no one was killed, but the damage was estimated at $35 million, making it one of the most expensive fires in New Zealand history.

Fire risk is an active area of research at the University of Canterbury and at Crown research institute Scion. At both places, researchers model how fires spread, with the aim being to better understand how we might reduce our vulnerability to increasing fires, for instance through thoughtful land use or improved fire-fighting strategies.

More fires aren't just bad news for us; they also have the potential to spell catastrophe for our forests and grassland. According to Dr Barbara Anderson, an ecologist and Royal Society Rutherford Discovery Fellow at Otago Museum, our native bush was quite fire-resistant back when it was intact, because it was damp and dark, but the fact it's now fragmented and also didn't evolve with regular burning means it's become particularly vulnerable to fires. Australia's flora, by comparison, has adapted to deal with regular

167

wildfires, although the massive Black Summer fires caused devastation on a scale not witnessed before.

If fires are powerful enough, they can actually generate their own weather – including one of the most apocalyptic-looking phenomena going, the firenado. For starters, fires of a certain size can generate their own clouds, known as pyrocumulus because they're convective clouds (cumulus) that form when buoyant air rises. (Fires provide plenty of very buoyant air, and plenty of particles for cloud-water drops to form on to.) When a cumulus cloud is large enough and the updraughts in it are powerful enough, a cumulonimbus (rain cloud) results. Cumulonimbus clouds are stormy characters, often carrying heavy rain, hail, thunder and lightning.

That means the chain reaction of fire-generated clouds can go something like this: an intense fire creates pyrocumulus clouds, which turn into cumulonimbus clouds, which go on to generate their own severe weather, which may even include lightning strikes that could go on to start new fires. And sometimes, a pyrocumulus might draw the flames burning below up into itself, possibly resulting in a firenado just like the sky-high red-hot licks sighted during the California and Australia fire seasons over the last few years, and at least once during the Port Hills fire here in 2017.

Furthermore, fires also add an unwanted extra injection of carbon dioxide into the atmosphere. As mentioned, the Australian Black Summer fires of 2019–20 are estimated to have put more carbon dioxide into the atmosphere than a whole year of Australia's human emissions. During El Niño conditions, regions in East Asia, Indonesia and northern Australia become drier and can experience severe drought conditions, leading to increased fire activity. In other words, it can become a vicious fiery cycle: climate change causes more fires, which emit more carbon dioxide, which causes more climate change.

Fire is one of the main considerations when it comes to planting trees to soak up carbon dioxide and slow climate change. While tree-planting is a powerful way to offset emissions, which is necessary to reduce the chance of fires, it's also important to be careful not to add fuel to potential fires – remember that warmer and drier conditions mean trees and other fuels for forest fires dry out faster, and stay dry for longer. As part of the One Billion Trees programme, the New Zealand Government has set the goal of planting a billion trees by 2028, with the motto 'Right tree, right place, right purpose'. This means tree-planting must take account of both current and future fire danger, the susceptibility of the trees to burning, and the availability of water.

We know from first-hand experience during Australia's Black Summer that the smoke and particulate clouds associated with major fires can be transported far from their source regions. This is yet another way that fires can directly affect the weather and the climate. The smoke plume from the Australian fires, for instance, was estimated to be as powerful as a moderate volcanic eruption in terms of blocking out sunlight, for a short time at least. In isolation, the smoke plumes from increased fire activity would have a slightly cooling effect worldwide, but the increase in greenhouse gas emissions outweigh this, and last much longer in the atmosphere.

*

Fires are not the only way that New Zealand is set to feel the heat in the future. As the climate warms, the number of days we would call 'very warm' or 'hot' is growing. A good rule of thumb is that, with 2°C of average warming, the number of hot days for any given area – say, 25°C for Wellington, 30°C for Christchurch and 20°C for Invercargill – roughly triples.

With another degree or more of warming, what was once a rarity will become commonplace. The years 2016 and 2020 are the warmest on record globally, at the time of writing in 2022 at least. To put that another way, no year in the historical record – going back 140 years or more – has been warmer than 2016 or 2020. We have had a little over a degree of global warming in those 140 years. If we get another degree of warming by the end of this century, it's virtually guaranteed that every year after about 2035 will be warmer than 2016. That would mean that, by the mid-2030s, the global climate would be outside anything seen for several thousand years. And, as the climate continues to warm, the ratio of heat records to cold records is expected to reach about 20:1 by mid-century, and 50:1 by the end of the century.

Furthermore, with 2°C of average warming, the marine heatwave conditions experienced here in New Zealand during the summer of 2016–17 would become typical. A heatwave on top of that would be completely outside our collective experience.

It's worth taking a moment here to remind ourselves that, in fact, today's climate is already outside the experience of those who were alive in 1900, or even in 1950. One of this century's 'colder' years in the global average, 2004, would have been counted as off-the-scale hot any time earlier than about 1970. That's how quickly temperatures are rising – and it also tells us how short our memories are.

The other thing it tells us is that it's only a matter of time until the temperature in Canterbury hits 43°C or higher. How long exactly is really up to us.

*

I live on the Kāpiti Coast, a few hundred metres from the shoreline. It is a five-minute walk to the beach, where a row of houses are built along the top of sand dunes, directly above the beach. Many sit five or more metres above sea level, but their front lawns start only a few metres back from the high-tide line. It's a strange and rather upsetting feeling to walk past my neighbours' homes and know that, as sea levels continue to rise and onshore winds drive waves inland, their front lawns will begin to be eaten away, followed by the houses themselves.

Thousands of homes just like these sit within mere metres of the New Zealand coastline, as do billions of dollars' worth of roading and other infrastructure. As sea levels rise, all of these places become increasingly exposed. The question is: just how much could sea levels ultimately rise?

So far, sea levels around the country have already risen around 20 centimetres in the past 100 years, and we're guaranteed to get at least 30 centimetres more over the next

40 to 50 years, reflecting the acceleration we're already seeing in the rate of sea-level rise. In many parts of the country, that's enough change to make what was once a 1-in-100 year erosion event a 1-in-4-year event.

Over the next century or two, our local sea levels are expected to rise a little faster than the global average. Half a metre of global average sea-level rise would equate to about 55 centimetres around our coasts. Beyond the next 50 years, it depends very much on how quickly greenhouse gas emissions are reduced. If we act now, and if we're lucky, there may be less than a metre of sea-level rise over the next century; if we don't take any action, and if we're unlucky, sea levels may rise by nearly two metres in that same timeframe.

The real trouble is that if we get onto that high track, there is no stopping the melt of a lot of the ice in Antarctica and Greenland, with many more metres of sea-level rise in store for centuries into the future. The West Antarctic Ice Sheet, which sits on the Antarctic continent between the Ross Sea and the Weddell Sea, is one of our main concerns. If all of it melted, that would add between three and four metres to sea levels – and, while fully melting the whole ice sheet would take hundreds of years, the worry is that unstoppable melting could set in some time in the next 20 or 30 years. The ice is so heavy that the land it sits on has been pushed down below sea level, meaning that warming ocean water could get over the coastline and in and under the ice. If that happens, the warmed ocean water will run downhill and start to float whole glaciers and large parts of the ice sheet. If that process starts, it won't be possible to stop it.

We already know from the geological record that in the past, when carbon dioxide levels were about as high as they are now and temperatures only a degree or so warmer, sea levels were several metres higher than they are today. We understand that at some level of warming, we'll lock in the runaway melting of the West Antarctic Ice Sheet. What that level of warming is, and exactly how the feedbacks and instabilities will play out, is very difficult to estimate. It remains an area of deep uncertainty in our understanding of the future. However, we do know that, if the planet keeps warming, it will become inevitable at some point – at an educated guess, maybe at 2°C of global warming? When and if that happens, we can be sure parts of the Greenland ice sheet will also be melting rapidly. In addition, parts of the East Antarctic seem to be vulnerable to the same runaway melting as in the west of the icy continent, and that could add several metres more to sea levels. If global warming does get over that threshold (2°C, perhaps), we know we'll lock in at least five metres of sea-level rise – and maybe even double or triple that. There were some periods, while Earth was emerging from the last ice age, where global sea levels rose four metres in a century. That's a metre every 25 years. Perhaps that won't happen in the future, but we do know it has happened in the past, so it's not impossible.

Aside from the possible thresholds for irreversible melting, we know that sea levels will continue to rise for centuries, because it'll take the oceans and the ice sheets a long time to adjust to the changes we have already made to the climate. The latest IPCC Assessment Report (the sixth, released in 2021) stated that even in a Paris Agreement future – where we get to zero emissions of carbon dioxide by 2050 and we limit warming to 1.5°C or only a little more – we could see up to three metres of sea-level rise by 2300, as ocean waters continue to expand and as ice continues to slowly melt.

If all of the ice currently on land melted and flowed into the sea, global sea levels would rise by around 70 metres. Granted, melting all the ice on Earth would take thousands of years to play out – but, once again, we could lock in a lot of that melting before the end of this century. The Earth has been there before. It's a change that's difficult to fathom – it would be far beyond anything humanity has ever had to face, for all of recorded history and beyond. Large swathes of coastal land in Asia, Africa and the Americas would be submerged. Auckland, Hamilton, Palmerston North and many more towns would disappear underwater. But the real question if that happens isn't which cities or continents will disappear; it's whether billions of humans can even survive in such a changed world.

Whatever happens, if we do cross 2°C of warming and lock in at least five metres of sea-level rise, it will change the map of the world for thousands of years into the future. Many low-lying islands will disappear completely, and regions that are close to sea level – downtown Auckland, South Dunedin and Lower Hutt, to name but a few – would be largely underwater. Globally, hundreds of millions of people would have to move.

To me, it doesn't matter that the melting itself would take hundreds of years to play out. What matters is that we could lock it in within just three or four decades. That's an awesome responsibility to face up to and take on.

*

If Kiwis aren't living near the coast and staring down the threat of sea-level rise, they're in the path of a potential river flood, with around two-thirds us residing on flood plains.

The way heavy rainfall is changing will lead to our rivers flooding more frequently, and occasionally at record-breaking scales. In February 2004, a major storm caused the Hutt River to flood. It was summer, but the storm was very like a wintertime event. The different season meant sea temperatures several degrees warmer than in winter, and in that sense the storm may be an indication of the sort of event that will occur more often in a warmer future. One study of the Hutt River catchment showed that, with 2°C of warming, a 1-in-100-year flood would occur about twice as often. The flood would inundate large sections of Lower Hutt and Petone, affecting thousands of people and likely costing hundreds of millions of dollars in damage. If emissions of greenhouse gases are not reduced and we get to 3 or 4°C of warming, that sort of flood would occur around five times as often. The very biggest floods would likely increase in volume by 30 to 60 per cent, depending on the level of global warming.

The Canterbury Plains are covered with big braided rivers that stretch across the landscape like a network of veins. These rivers are so shaped because they meander, changing course quite quickly between their current banks. Over longer timeframes, they can change course dramatically, and the river mouth can even move many kilometres up or down the coast. For instance, the Waimakariri River has its mouth near Kaiapoi, north of Christchurch, and that is where it has stayed in the time that the city of Christchurch has developed. But, in the past, the Waimakariri has flowed right through the area where Christchurch now sits; at times, it has even flowed to the south of Banks Peninsula, into Lake Ellesmere/Te Waihora. Naturally, there has been a lot of work over the past century to build up the current banks and reduce the risk of such a dramatic change in the course of the river. Whether or not some future super-flood over-rides the stop banks, only time will tell.

The major rivers that flow east from the Southern Alps are fed by rain and snow from the mountains, and by glacier melt. As climate change causes more glacier ice to melt, river flows will generally increase – but, once a glacier becomes too small, it will no longer contribute much to river flow. The meltwater from some small glaciers around the world is already declining, so it seems that the smaller the glacier, the quicker it reaches its peak of melting. Bigger glaciers on the other hand, such as those in the Himalaya, will go many decades yet before their contribution to river flows starts to wane. Here in New Zealand, glacier melt is well under way, though there is still a lot of ice in the Southern Alps. Peak flow into South Island rivers is some years away yet, but our glaciers are losing ice rapidly.

When a southern winter sees more snowfall than rain, the snow builds up in the mountains and the main mountain-fed rivers – such as the Clutha/Mata-Au, the Waitaki and the Waimakariri – have less water flowing in them. Come spring and summer, a lot of that snow melts, raising river levels substantially. By contrast, if the winter features more rain, the river will be fuller during winter. Then, when spring comes, there'll be less stored snow to melt and the river levels won't rise so much.

At present, this difference in river flows between summer and winter is a bit of a headache for hydroelectricity generation. Winter, when everyone is heating their homes and staying indoors, is the time of strongest power demand, but it's also the time of the lowest inflow to the big hydro lakes. Meanwhile, summer sees weaker demand, right when the lakes are busy filling up. So, in a way, climate change makes hydro generators' jobs easier, as it raises flows and lake levels when demand is higher in winter, and reduces flows when demand is lower in summer. At the same time, seasonal demand itself is changing. In a warmer climate, people need less winter heating and more summer cooling, acting to equalise electricity demand at the same time that electricity supply is becoming more even throughout the year.

However, climate change isn't all good news for hydroelectricity in New Zealand. As the climate warms, river flows also become more variable. While average flows are on the rise in winter, the chances of very low flows and very high flows are also increasing. Even in a warmer climate, we may still see years where winter hydro-storage becomes marginal. At the other end of the scale, river flows may be so high at times that water has to be spilled from dams – which, to companies that generate hydroelectricity, is literally throwing money away.

*

One class of rivers expected to only get bigger are ones that we cannot even see. An atmospheric river is a great plume of water vapour that usually extends from the tropics to the middle latitudes, but can exist anywhere there is a moisture source and winds to carry it long distances. While it has long been known that water-vapour transport in the atmosphere is a vital part of the climate system, it's only in the last quarter of a century that these long filaments of moisture – these rivers in the sky – have been studied closely.

A single atmospheric river can be several thousand kilometres long, and can carry more water (in the form of gas) than the Amazon. Atmospheric rivers can be rated according to how much moisture they carry, from a weak category one that brings 'mostly beneficial' rains to the strongest category five, with heavy, 'mostly hazardous' rains. And, when an atmospheric river makes landfall, the results can indeed be catastrophic. The west coast of the US is perfectly placed to be on the receiving end of massive atmospheric river flows, with one category four or five event somewhere from California to Washington state every couple of years or so. Even though they are episodic, these events account for about one-third of the total rain and snow that falls along the western US coast, and the flood damage averages around US$1 billion per year.

New Zealand is similarly well placed for atmospheric rivers that come from the northwest, from over the seas to the east of Australia. Every major flood in the Waitaki River between 1979 and 2012 has been the result of atmospheric river activity, and the loss and gain of ice in South Island glaciers is clearly linked to atmospheric rivers. When the moisture from these flows falls as snow, the glaciers gain mass; when it falls as rain, they lose mass. As the climate warms, the expectation is that atmospheric rivers will deliver more rain than snow to the Southern Alps, causing the South Island glaciers to shrink further and adding even more water to river flows.

Atmospheric rivers featured several times in the flooding that struck New Zealand in July and August 2022. They brought huge amounts of moisture out of the tropics and down over the country and, in between, subtropical storms, thunderstorms and vigorous fronts from the Southern Oceans battered the country. One atmospheric river event in mid-July, in particular, affected the northern South Island and much of the North Island, flooding roads, and causing slips and power cuts that affected thousands of people. Another during August led to the State of Emergency declared in Nelson, Tasman, the West Coast and Marlborough. That one was estimated to be the strongest and wettest August atmospheric river for over 60 years. Then, in January 2023, ex-Tropical Cyclone Hale and its associated atmospheric river brought torrential rain and flooding to the Coromandel and the eastern North Island. Later that month, another atmospheric river brought unprecedented rain to Auckland with widespread flooding, infrastructure damage, and tragically, the loss of four lives.

As the climate warms and the amount of moisture in the atmosphere increases overall, atmospheric rivers are getting bigger, in terms of the amount of moisture they transport. They are also expected to become wider and more persistent, leading to larger rainfalls over wider areas. As well as resulting in overly full hydro dams in the South Island, more powerful atmospheric rivers could bring very damaging flooding, especially to western areas of the country, something we already saw happen in winter 2022.

*

The increased amount of moisture in the air is also a source of energy for storms. There are a number of reasons a warmer climate might be expected to bring stronger, more vigorous storms to New Zealand, and that is broadly what we are seeing – but what we don't see is more storms in total. We'll definitely have stronger storms, but if anything there'll be fewer of them. However, thanks to the fact that the winds and rains will be more intense, storm damage is still likely to increase.

In terms of winds more generally, the expectation is that the westerlies that blow over the country will gradually get stronger over the coming century.

One of the main stormy factors here in Aotearoa is the negative phase of the Southern Annular Mode (SAM), a classic of which occurred during the 1957–58 summer. It was brutal. Storminess was common across the whole of the country, but the worst weather occurred over the southern half of the South Island. Flooding was extensive and frequent, making Christmas pretty miserable for many in Southland and South Otago. The Clutha River/Mata-Au had its highest annual flow on record by a factor of two – a massive amount of water that has not been matched in the six decades since.

As well as the SAM being extremely negative during the last three months of 1957, that summer was also a season of El Niño conditions, which would have helped bring the unsettled weather to New Zealand.

But could that happen again? In short, yes.

A prolonged period of stormy conditions and negative SAM is always on the cards. The atmosphere is now around 1°C warmer than it was in the 1950s, and there is around 6 per cent more moisture in the air on average. So, if we did see a prolonged period of negative SAM in future, it could be even wetter and more unpleasant than what people experienced at the end of 1957. On the flipside, the positive trend in the SAM since the 1950s means we're also less likely now to see such a long spell of the negative SAM. So, fingers crossed.

There was one silver lining to the terrible weather over New Zealand that summer. When the SAM brings stormy weather to New Zealand, the skies near the Antarctic coast are generally clearer. In December 1957, a message came through from Scott Base to report that the New Zealand Antarctic Expedition tractor party – including none other than Sir Edmund Hillary – were making excellent progress in 'brilliantly fine' weather. It seems that the negative SAM at least helped Hillary and his party on their way to the South Pole!

*

In the winter of 2011, cold air outbreaks brought snow to much of the country. It settled for a day or so around Wellington, and a few flurries even made it as far north as Auckland. I stood by the window, watching the snow falling in my backyard and thinking this would be a once-in-a-lifetime experience.

Two things are required for a snow storm: moisture and freezing temperatures. But, the way the climate is changing, those two things act to cancel each other out. A warmer atmosphere holds more moisture, but is less likely to get down to sub-zero temperatures. A moister atmosphere might hold the potential for heavier snow, but it has to be cold enough, especially near ground level. Cold blasts resulting in heavy snow are still possible over parts of the South Island, such as were experienced in 2022 during September and October but, as time goes on, cold enough air flows for snow will be harder and harder to come by, especially north of Cook Strait. It is possible I'll see snow again in Kāpiti, but I'm not holding my breath.

Snow will keep falling in the Southern Alps for a long time yet – in fact, they might even get more seasonal snow in winter, at least at high elevations. That's because, if you go up high enough, even in a warmer world, the temperatures will still be below freezing – and, if the air is carrying a bigger freight of moisture, that should translate to more snow in places that are cold enough. However, the overall global warming is still likely to win out. This idea was tested by NIWA in a series of model simulations that allowed for various degrees of warming, and showed that, even at the tippy top of the Southern Alps, snow mass is still projected to decrease over time.

In the North Island, increasing temperatures may preclude snow falling in all but the coldest of cold outbreaks. However, at higher elevations – over the Volcanic Plateau and in the mountains – snow will likely remain for decades yet.

One of the most obvious victims of the way climate change is affecting snowfall over the country is our ski fields. In the coming decades, as the climate warms, the ski season will become ever shorter, starting later and ending earlier. Generally speaking, the farther north you go and the lower your altitude, the sooner there will be trouble. Ski fields in colder places like Central Otago may be fine for decades to come, while those in the central North Island could be marginal in 20 or 30 years. As the climate warms, cold-weather activities like skiing will become less possible for everyone, everywhere.

We're already seeing this play out with what's happened recently with Ruapehu's ski areas, and change is on the cards in the way other ski fields are adapting. Winter 2022, for instance, might have brought bumper snowfalls to many of the South Island ski fields, but these will become rarer and less predictable in future, even around Queenstown. With that in mind, Cardrona Alpine Resort, near Wānaka, has already moved to incorporate a range of summer activities in its offering, putting ski lifts to use in summer carrying mountain-bikers and their wheels up the slopes, and changing ski runs to adventure trails. That kind of thinking is what's needed to keep ski operators going, as the 'ski' part of things becomes less reliable. Fields will need to adapt, and the fact some are already doing so is telling. As with everything related to climate change, it's not so much a matter of if as when.

As for that other icy white winter phenomenon, it's actually possible – even in a warming climate – that the number of frosts in some locations could go up. This would happen, for a while at least, if the pattern of winds and clouds changed in such a way as to give a location clearer skies and lighter winds in the winter. Even if that happened, however, it would be very much the exception. Over most of New Zealand, the number of frosts has already decreased by a half or more since the middle of the twentieth century. With another degree of average warming, frost will become a rarity in most of the places we live in New Zealand. At that level of warming, or beyond, we would have to go into the mountains if we wanted to see frost on the ground.

Whatever the rate of warming, we know that change is in the forecast for New Zealand. How much remains to be seen. But, as summers become longer and hotter, as winters wither, as alpine snow and ice recede, our country is going to get a complete makeover. It won't be the same. The more warming there is, the greater the change will be.

We will grow used to the changes, of course – what other choice do we have? – but the country we call Aotearoa will be a very different place in a world that is even a degree or two warmer.

10

The Pacific's outlook

Beyond our shores, but still in our big blue neighbourhood, the forecast for a warmer climate contains many of the same features – but, for the Pacific, much of what's to come is very extreme. As well as rising sea levels, there'll be stronger storms, more rain, more floods. Furthermore, the countries of the tropical Pacific have fewer resources than we do to defend themselves.

Naturally, things are also going to get hotter and, since the tropics are where temperatures vary the least, things are going to shoot off the charts much more quickly than here in New Zealand. Across a lot of the tropical southwest Pacific, temperatures usually vary from year to year by no more than half a degree, so warming of 1.5°C would be enough push things beyond previously recorded extremes. At that rate of warming, most of the southwest Pacific would be experiencing high-temperature extremes beyond anything local populations are adapted to by the middle of this century, just 30 years from now.

During the wet season, from November to April, the southwest Pacific is also going to experience an increase in rainfall variability. In other words, how much rain comes and when it comes is going to change – at times, some parts of the region will have more rain than they can handle, while others won't have enough. Here, it matters where countries sit in relation to the South Pacific Convergence Zone (SPCZ) – that band of intense cloudiness and rainfall that stretches southeast from Papua New Guinea towards French Polynesia, running out of puff when it merges with weather systems over the central South Pacific. Variability in rainfall is expected to increase as variability in the SPCZ itself increases. But, for countries sitting on the southern side of the SPCZ, variations are likely to be weighted towards drier years, at the expense of wet years – and that would bring a big increase in more severe droughts. Meanwhile, for countries on the northern side of the SPCZ, variations are likely to favour wet years, at the expense of the dry. In the climate of the past century, countries in the northern part of the region might have seen double their normal annual rainfall reasonably often; in a world that's 2°C warmer, those same countries may see annual rainfalls of five or more times normal every few years.

Rainfall variability across the tropical southwest Pacific is going to change, and with it access to freshwater. Especially for countries in the southern part of the region, being able to store freshwater when it's available and manage it in between rainfalls will only become more critical than it already is. For countries closer to the Equator, flood management is going to be the challenge, in some years at least.

Rising sea levels only add to this need to protect freshwater supplies. In coastal zones, the water table – the depth at which groundwater sits – rises and falls with the sea. Freshwater below the land surface feels the pressure of seawater just offshore and, as sea levels rise, so does the water table. This means that, as sea levels rise, salty seawater begins to penetrate underground, across the beach zone and into fresh groundwater near the coast. That results in salination, turning previously useful and drinkable water supplies into salt water no good for agriculture or drinking. Saltwater from the ocean is already infiltrating the fresh groundwater lenses that a lot of Pacific islands depend on for drinking water and for agriculture. In some places, crops are already being planted in large raised beds in order to keep the plants from coming in contact with salty water.

In terms of tropical cyclones, the changing climate is a bit of a mixed bag. It all depends on which ocean basin you're looking at, but there are a few consistent patterns. Across all tropical ocean basins, total tropical cyclone numbers are expected to decrease a little, while the maximum intensity of storms increases. The clearest signal is that tropical cyclones will produce more intense rainfalls. This is a direct consequence of there being more moisture in a warmer atmosphere – whenever there's a storm, wherever it is on the globe, more water will fall out of the sky.

On average, around ten tropical cyclones form across the southwest Pacific each cyclone season. By the end of the century, that number may have decreased to nine, or possibly eight. But these storms will be more likely to be intense and likely moving more slowly and producing more rain. That's a recipe for increased cyclone damage across the southwest Pacific. A recent example of such a slow-moving, intensely rainy and devastating storm was Hurricane Harvey, which sat near the coast of Texas for several days in 2017 and dumped around a metre of rain in the Houston area and elsewhere. The extensive flooding and flood damage led Harvey to be one of the costliest tropical cyclones ever recorded – that single storm is estimated to have cost the US economy around US$90 billion. (And an attribution study carried out by New Zealand and US scientists suggested that human-caused climate change was responsible for around US$60 billion of that damage.)

The most intense tropical cyclone so far recorded in the Southern Hemisphere, and the costliest in terms of damage to property, was super cyclone Winston, which hit Fiji and Tonga in 2016, causing extensive damage and claiming 44 lives. Like Harvey, Winston travelled relatively slowly when it was most intense, looping back on itself and changing course several times.

Storms like Winston and Harvey are what we expect to see more of across all tropical ocean basins in future.

*

One of the most important parts of the Pacific's forecast for a changed climate is its impact on food and water. If temperatures in the region are pushed to new extremes, they will be beyond what local food crops are adapted to. Through a combination of unprecedented temperature extremes and more variable rainfall, risks to food security are likely to increase substantially this century.

Fish are a fundamental part of the diet across the Pacific, and fish exports form a significant part of many countries' economies. As temperatures rise, vital fish stocks will be forced to migrate and the coral reefs that ring many Pacific islands will decline or, worse, disappear. With 2°C of warming and ongoing acidification of ocean waters, it is expected that all tropical corals will die out. The effects on reef-based marine ecosystems would be profound, not to mention the psychological impact of the death of such an iconic part of the marine environment of many Pacific Island states.

As the century progresses, changes in the climate and in the oceans across the southwest Pacific will collectively make life and livelihoods harder and harder to sustain. Some island states will, most likely, be forced into deciding whether or not it's time to pack up and leave for good. Which nation will be the first is hard to say, but we may well find out before the middle of this century.

As time goes on, more and more people will be displaced – and I believe New Zealand has a moral obligation to extend our support to our Pacific neighbours as climate change continues to bite. We must do what we can to support nearby island nations with disaster recovery. In my opinion, it is our duty to provide assistance when it comes to adapting to a warmer future, to sea-level rise, to changing agricultural yields and fish stocks, and to water-storage requirements.

Most importantly, we can – and I believe we must – support any displaced communities that need to find a new home in New Zealand. We must open our arms and our doors to all Pacific nations. That's what good neighbours do.

11

Beyond the weather

A warmer world will bring profound changes to our weather systems, but the changes will extend beyond exactly what we see in the sky, and what falls out of it. It will change the look of the land around us, and what grows and lives on it. Since our lives – and those of every other living thing on the planet – are inextricably connected to the conditions of our environment, it really goes without saying that changing the weather means changing life as we know it. That will be bound to change how we think about our world, our whenua.

There are going to be many knock-on effects, and we are already seeing many of them play out. Floods and droughts and changes in temperature put food and water security at risk, and that only adds to global conflicts and instability. Most ecosystems are not adapted to repeated heatwaves, and as temperatures rise, plants and animals struggle to survive – and that includes fish and other aquatic species living in rivers and lakes, and marine life in the upper ocean.

There are also significant cultural implications to the various impacts climate change will have on our flora and fauna. For centuries, iwi and hāpu have accumulated and passed on critical local knowledge about when to gather food and plant crops. As the climate changes, so too does the way things grow, meaning this centuries-old knowledge will similarly be forced to change, and much of it will no longer be relevant. Some practices may decline or even disappear due to environmental change.

Human infrastructure also feels the heat, with power lines sagging and causing shorting and brown-outs in power networks. At the same time, power demand spikes when it's hot and everyone wants their air-conditioning on. Furthermore, in hot and dry conditions, sparking from power lines onto dry trees can set off major fires, such as those seen in California in recent years.

When it gets hot, roads surfaces tend to melt. This can cause traffic accidents, and disrupt road transportation. Heat can buckle railway tracks and cause delays or complete stoppages in train services. Even now, warm summer days in the Wellington region can result in go-slow conditions being imposed on commuter-train services because the tracks have warped in the heat.

Air services, too, can be badly affected by heat: as air warms, it becomes less dense, and if the density decreases too much aircraft can't generate enough lift to get off the ground. They may end up having to offload freight, passengers, fuel or all three in order to get airborne.

*

Here in Aotearoa New Zealand, our native flora and fauna already face a number of threats; climate change is just one more that they certainly don't need.

Across the globe, we are already seeing how climate change is only adding to the other pressures humanity is placing on the natural world. As we have expanded across the planet – in terms of our numbers, the land we take up and the resources we use – the ecosystems that Earth's flora and fauna rely on for survival have receded and become fragmented. This has massively reduced biodiversity, and driven some species to extinction or the brink of it. Species are now going extinct hundreds of times faster than before we came along, and this die-off is known as the sixth extinction – the latest in a series of mass-extinction events the Earth has experienced over the last half-a-billion years. The most recent one – the fifth – involved the impact of a giant asteroid on the coast of Mexico and the extinction of the dinosaurs. This time, we are the asteroid.

There's an awful lot we don't yet know about how climate change is affecting New Zealand's native species, or how it will change the dynamics of our natural environment, but ecologists such as Dr Barbara Anderson (the Rutherford Discovery Fellow at Otago Museum I mentioned in an earlier chapter) are busy working on filling the gaps in our knowledge with their research. And, even if we don't yet have all the information we need, many conservationists and scientists like Barbara are acutely aware of how quickly we need to act, and frustrated by how little is being done to prepare for what lies ahead. The way Barbara explains it, 'There's this attitude that New Zealand will be OK – it has a different climate, it's an island. But we shouldn't be waiting until we see the effects of climate change on our country's species before we do something about it.' Top of the agenda, according to Barbara, should be making our biodiversity more resilient to climate change through better land-use, and by more effectively allocating the resources available to maintain our conservation estate, where the bulk of our taonga species reside.

One thing we can be fairly sure of is that, as the climate warms, New Zealand and its surrounding waters are going to become more hospitable to a range of new and destructive pests. One example is myrtle rust, a serious fungal disease that was first detected here in 2017 and has spread rapidly since. The most likely explanation for how it got here is on the wind from Australia, where it was first found in 2010. The spores are carried on the breeze, and when it arrived in New Zealand myrtle rust found an abundance of plants in the myrtle family to attack – including many of our taonga species, notably pōhutukawa, mānuka and rātā. It forms bright yellow-orange powdery pustules on the leaves, fruits and flowers of the plants it infects, and it cannot be eradicated from Aotearoa. With climate change causing weather patterns to alter and winds to become stronger, it's more than likely we could see other such arrivals with increasing frequency in the future.

Part of the reason our native plant species are so vulnerable to climate change and the new threats it could introduce is that, over the course of human settlement in New Zealand, their natural habitats have become fragmented. Before Polynesian and European colonisation, New Zealand was almost completely covered in native forest – but, when humans came along, we started felling trees for settlements, agriculture and transport, and over millennia managed to destroy nearly three-quarters of the natural forest cover. Now, native bush exists in discrete parcels, with vast swathes of unforested land in between. It's very different from how it once was.

Kauri dieback provides an example of how climate change can add to the pressure already being felt. Like myrtle rust, kauri dieback cannot be eradicated from New Zealand, and is a critical threat – the microscopic fungus-like organism infects and damages the tree's roots, preventing it from being able to draw the water and nutrients it needs from the soil, so the tree eventually dies. Since we can't get rid of it, the only thing we can do is try to halt the spread by avoiding transporting infected soil through forests on our clothes, footwear, equipment and vehicles. And sure, kauri dieback is caused by an introduced pathogen, not by climate change – but, as Barbara points out, 'Why is that pathogen suddenly taking off now? The trees are fragmented. They are also suffering from drought. Kauri dieback is really the final nail in the coffin.'

*

There's an unassuming little creature here in Aotearoa whose ancestors have been around since long before our current series of ice ages began. It is, of course, the tuatara – the spiny and lethargic reptile that can grow up to half a metre long and still looks a lot like its ancient ancestors, which lived in the time of the dinosaurs, 200 million years ago. In all that time, the tuatara has barely changed, simply because it hasn't really needed to – but then we came along.

Habitat loss and introduced pests, such as rats and mice, devastated tuatara populations, and nowadays they exist only on island sanctuaries and in a few protected pockets on the mainland. From my office at Victoria University, Zealandia wildlife sanctuary is just a short bike ride away (yes, I still get around on my bike a lot of the time, just as I did when I was a kid – though these days it's an e-bike!). There, I can visit the tuatara that often hang out on a crumbling bank conveniently close to one of the main paths.

As well as being incredibly slow-growing, tuatara are very slow-breeding: they only get started when they reach the age of 15, and females breed only every nine years or so. That means that every single egg that hatches into a healthy baby is a major win for the species.

Tuatara eggs also happen to be incredibly susceptible to temperature change. As the eggs warm, the gender of the hatchlings within can alter. With hotter conditions, more males are likely to be produced. Over a decade ago, a group of New Zealand scientists studied the rare species of tuatara (*Sphenodon guntheri*) that lives on North Brother Island in Cook Strait, and observed that there was a 60 per cent skew towards males in the population. The scientists then simulated how long the island's existing tuatara population was likely to persist, based on varying percentages of males. If the sex ratio reached up to 75 per cent male, the population was likely to persist for at least 2,000 years. If the male population got to 85 per cent – something that's possible with a warming climate – the population would be extinct within around 300 years (or eight generations). When you add the problem of too many males to the existing threats posed by habitat loss and pests, you see clearly how climate change could obliterate a species already on the brink.

Tuatara are not alone in their vulnerability to a fast-changing climate. All of our species are going to face their own reckoning with climate change, just as we are. Aotearoa is widely recognised as a biodiversity hotspot, home to around 80,000 endemic species, but ecologists fear that even modest temperature increases could push many of them over the edge. In Barbara's words, 'If you've got a fragmented habitat and the climate is changing, it's harder for a species to disperse to the next place.' A number of our native species have already been forced to migrate to survive. The kākāpō, one of the world's most endangered species, was not originally an alpine bird, but fled to mountainous areas of the country to get away from humans and invasive pests. Like the tuatara, the kākāpō now only exists in island sanctuaries.

*

If hot days can be hard for people and native animals to cope with, they can be even more testing for the crops and creatures we rely on for food. Dairy cows experience heat stress when it gets too hot, and that's only going to become more of a problem as things get warmer. In Northland at least, the suggestion has already been made to keep cows indoors during the warmest months of the year, in order to avoid undue heat stress.

Crops also suffer both in extreme high temperatures and from the associated lack of water that comes with drought. These two things in combination can be devastating for crops and natural ecosystems alike. One way to measure what plants can grow where is by looking at what's called heat units – numbers that represent how much total heating a plant receives over a growing season. Calculating the heat units for a given plant is pretty straightforward.

- First, find the base temperature. This is the temperature below which the plant would not grow at all – for a lot of crops in New Zealand, that's often 10°C, maybe 5°C.
- Then, for each day of the growing season, work out the average temperature. For instance, if yesterday's maximum was 20°C and the minimum was 10°C, then the average temperature was 15°C.
- Finally, find the difference between the average temperature and the base temperature – that's your heat units for that day. For instance, in the example above, the difference is five – so you have five heat units above a base temperature of 10°C.

To find the total for the season, you just add up the daily heat units. Certain crops require particular totals of heat units to reach maturity. For instance, oats need about 1,600 heat units to reach maturity, while corn needs over 2,000. By looking at the numbers around the country, we can see where certain crops grow now and where we might be able to grow them in future.

If we get another degree or so of warming, total heat units through the rest of the century are set to increase by about half. That would mean regions such as the Wairarapa and

Canterbury would have the same kind of numbers currently seen in the Auckland region. Crops currently grown in the southern North Island would become viable in Southland. Meanwhile, in Northland, bananas would become a staple crop – and, in fact, bananas are already growing happily in Northland, with farming them under discussion. Pineapples would likely become an option in a few decades. This is the pattern we'll see over the whole country: as things get warmer, the crops we grow will migrate south, and new subtropical crops will come into the north.

All this tropical fruit might sound quite marvellous, but there's a downside (of course there is): for some crops, it will be so warm they may not be able to grow in anywhere near as many parts of the country as they currently do. Some plants, especially fruit trees, need chilling in the winter to reset their clocks and tell them to fruit in the spring and summer. Apple trees need the most chilling, followed by apricots and peaches. Kiwifruit vines also need winter chilling to get them fruiting properly in the spring. The reason we get such wonderful apricots in Central Otago is because the trees get a proper dose of chilling in the winter – although the gold in the soil has to help!

As the climate warms, fruit trees will no longer get enough chilling in the winter. Chill units – the flipside to heat units –

are calculated in a similar way to heat units, but we measure how far *below* a threshold temperature we get. Chill units are usually done by the hour, rather than the day, but added up through the season in the same way as heat units. As the climate warms, and especially as we see big decreases in cold nights, chill units drop away rapidly. The Bay of Plenty will likely see decreasing kiwifruit production over time, with more grown farther south, as growers chase the chillier nights. Maybe Central Otago will remain cold enough in winter for decades to come, but Hawke's Bay and even Nelson might just become a little too warm in winter to support good fruit production in the spring and summer. There are chemical ways to trick the trees into producing buds and fruit, but those who want to do it the natural way will need to look south.

Natural ecosystems will do much the same as our food crops, migrating south as things warm up – at least, they will try to. Many will not be able to keep up with the pace of change. New plant species will become established where they couldn't grow before. Previously unknown plant pests are bound to flourish. Currently, kauri really only grow north of Hamilton (with one or two notable exceptions, such as those in the Wellington Botanic Gardens), but as temperatures rise it would be feasible for these regal trees to descend south across the country, with a little help.

Other plants and species will migrate upwards in search of the cold they require to grow, since temperature decreases with height. Many alpine plants are adapted to a particular range of temperatures and currently find them somewhere in the mountains of New Zealand. The only catch is that things are warming faster than plants can migrate – and, once you're at the top of a mountain, there's nowhere left to go.

*

New Zealand might be a long way from the rest of the world, but what happens in other countries on larger continents is going to have an effect on us too. We are very much part of the world. What happens to the world, and to the people who live in it, happens to us too.

One knock-on effect that we can count on, but not very adequately forecast, is what's going to happen between nations. As heatwaves and associated droughts become more common around the world, basic resources like food and water are going to become increasingly scarce in some places. People need these things – and they'll fight for them, if need be. What sort of conflict are we going to see in a warmer world, over resources, over land, over basic human rights no longer easily accessible to huge portions of the globe's population?

We've already witnessed the devastation droughts can cause internationally. From 2007 to 2010, the Middle East's Fertile Crescent experienced the worst drought on record. It caused widespread crop failure and a mass migration of farming families to urban centres, and there's evidence to suggest it contributed to the Syrian uprising in 2011. This sort of drought will only be seen more often in North Africa and southern Europe in future. It will put water security at risk for many millions of people, potentially triggering further conflicts in countries already struggling with civil unrest. Caring for the resulting refugees and migrants from these countries will a be one of this century's major humanitarian issues.

And where does New Zealand fit in all of that? Your guess, on this point, is as good as mine.

12

The response so far

Once you properly understand just how much climate change has already affected our world, and start to get a handle on how much it's set to affect things in future, it can feel pretty overwhelming. What can we possibly do in the face of so much change? How can we even begin to respond?

Well, for starters, it's important to remember that we're the ones driving all of this change in the first place. We might feel insignificant when we face up to the scale of what's happening, but we're absolutely not insignificant. We are, in fact, more significant than we might want to accept – because we're the ones who caused this mess.

The other thing that bears mention is that, actually, the response to this global problem is really quite simple: stop emitting greenhouse gases. Things only start to get complicated when other global issues – politics and commercial interests and 'the way things have always been' – get into the mix. It's those things, not the problem of climate change itself, that make it so difficult to know how to respond.

But, complicated or not, it's fundamental to hold on to the idea that there *is* a future to prepare for. The precise shape of our future is up to us, and always will be. We know what's happening, we know why, now we just need to work out a way forward, together. Already, we're seeing a number of responses take shape, in no small part because they have to, because our climate is already changing.

Knowing what's on the cards helps us to adapt to what might happen in future. There are many good reasons for us to make the adaptations we need to now, not later, because doing so will reduce harm to people, property and places. Our collective future can definitely be a lot cleaner, and greener, provided we set our minds to steering in that direction.

*

Here in New Zealand, many farmers are already navigating the challenges presented by more frequent and intense droughts in some places, and extreme rain events and floods in others. Farming is inextricably connected to the climate, whether it's crops or livestock, so it's hard not to notice the way things are changing. For a lot of farmers, adjusting to the climate of the future means making an effort right now to ensure what they do is as sustainable as possible, by reducing their emissions and restoring what's been depleted in recent decades.

You may have already heard of regenerative agriculture, the umbrella term for farming with a conservation and rehabilitation focus, where farmers focus on improving the soil and the water cycle. As well as building in resilience to climate change, the idea is to foster the long-term health of the land being farmed.

Family-owned Linnburn Station, in the heart of Central Otago, is a proud practitioner of regenerative agriculture. 'In 2014, we turned our back on the traditional soil management practices advocated by synthetic fertiliser and pesticide companies,' their website explains. 'We believe that healthy soil = healthy water = healthy animals = healthy people. Right now, we're on an amazing journey to see what works and what doesn't. We wish to openly share our experiences and our results with all those prepared to listen.'

Among the practices the station has implemented are cover crops, companion planting, zero-till (in other words, not disturbing the soil) and high-intensity grazing. These practices have replaced the use of synthetic fertiliser, fungicides and pesticides, with the station's end-goal being to improve soil health and increase soil carbon.

Peter Barrett, one of Linnburn's owners and the current family representative on the station, assumed responsibility for operations in 2012. He'd returned home from a successful business career overseas, and initially he simply wanted to make the farm a more financially viable enterprise. But, once he got stuck in, he saw there were bigger problems to address. In his words, 'We are in a brutal environment here, climate change or not. It's a very brittle place where we don't get constant rain every month. We get seven months without rain.'

They were spending oodles on conventional methods – planting clover and ryegrass, and applying fertiliser – but it was all for nothing if the rain didn't come. So, Barrett went in search of other ideas, and that led him to regenerative agriculture. The first thing the station tried was zero-till seeding. The way Barrett puts it, he was 'just taking bare seed and putting it in the ground' – and his efforts were soon rewarded. 'I started to get some success with it, and then I was away! Years ago, when you went out on the farm, you'd just see brown, summer or winter. Now you see green.'

As well as improving the health of the station's soil, the more sustainable practices have also proved kinder on the station's pocket. This is the sort of change that might not be easy, but it is necessary – and it is possible. As Barrett told *Country Calendar*, 'To change and step away from what you know is quite uncomfortable. But everything you can do conventionally, you can do regeneratively.'

*

Of course, it's more than our farming practices that are already being forced to change. Encroaching water – in the form of floods and sea level – is a major factor in a warmer future. First, floods: how do we deal with more of them, and ones that are bigger than have ever been observed?

One answer is to build levees or stop banks beside rivers. This has already been done for many of our major rivers, notably in downtown Lower Hutt, along the banks of the Hutt River, to help protect the CBD and residential properties. There is, however, a fairly hefty downside to stop banks: by giving us a false sense of security, they can actually end up encouraging more people to live and work right beside the river. Then, if a really big flood comes along – one that's beyond what the stop banks are designed to cope with – there can be even more damage to properties and more disruption to more lives than there would have been if the stop banks had never been built in the first place.

This is sometimes referred to as maladaptation – thinking you're adapting to climate change in a good way, when in fact you're just upping the danger levels and putting more people at risk. Another example are the sea walls that sit below beachfront properties. Like stop banks in a flood, sea walls can be over-topped in a big enough storm. And, as moisture builds in the atmosphere and sea levels continue to rise, most sea walls and stop banks are pretty much doomed to fail one day. They're both examples of wishful thinking – 'With just a bit of engineering, I can carry on with business as usual.'

The reality is different. River flooding, and flooding in general, will become more of a problem as time goes on.

Which brings us to the other possible response: retreat. In some cases, we really do need to move people and property away from flood plains and out of danger. Yes, doing so is painful and expensive, but it's the one thing that will ultimately give us, and our children, the greatest protection.

Taking climate change seriously is a vital ingredient in any effective response. As the world warms, it's already disrupting our lives, and will continue to do so. In a sense, it's better to go with the disruption you know.

*

I was once interviewed for television while standing on the beach not far from my home. At one point, the interviewer said to me, 'So, by the end of this century, all these houses we see along the beach here will be gone?'

It was a pretty confronting moment. Most of those houses have been there for decades. Each one is someone's home, and those people are part of my community.

I paused and took a breath, looking up and down the beach, before replying. 'Yes,' I said.

Sooner or later, homes that close to the shore will either have to be moved or they will be washed away. The reality is that, in many places, the coastlines are on the move, and we had better move with them. Some parts of the coast are rocky and fairly impervious to wave action, and other parts are actually growing as waves and currents deposit sediment. Other stretches already feature sea walls, which can be quite a good barrier and offer short-term protection, but the trouble is that all kinds of sea walls tend to get eaten away eventually. They are, at best, a temporary solution.

When a storm affects the coast, three things happen. First, the low pressure near the centre of the storm lifts the level of the sea surface up, typically about 20 centimetres in an average storm. The lower the air pressure and the stronger the storm, the more the sea level is raised. Second, if on-shore winds are blowing, they will pile up seawater at the coast, raising the sea level even more. Finally, wind-driven waves coming in to the beach will raise the sea level again. With just 10 or 20 centimetres of sea-level rise, the sea wall that was once just high enough to keep the waves at bay will find itself over-topped. I've seen just this happen near where I live in the last few years.

One of the country's real problem areas is South Dunedin, a residential suburb inland from St Kilda beach. As well as being very close to sea level, it is built on a combination of drained swamp and reclaimed land. For now, a very sturdy and tall sea wall along the nearby beach has kept storm surge and coastal erosion at bay, but that can't do anything to help combat the real problem: the height of the water table, or the depth at which ground water sits.

These days, the water table in South Dunedin is just below ground level, and the land is sinking slowly. The main problem with this is that, when it rains, the rainwater has nowhere to go. The stormwater system is largely ineffective because there is so little vertical room to move. In recent storms, fire engines were brought in to pump stormwater away because it just couldn't drain. That kind of problem is only going to get worse. In response, the Dunedin City Council and Otago Regional Council have established the South Dunedin Future programme, which aims to 'develop and deliver a climate change adaptation strategy for South Dunedin that works, is affordable, and that the community supports.' On the table in the past have been discussions about managed retreat, and building relocatable houses to take the place of the old villas that make up the neighbourhood.

The words 'managed retreat' roll off the tongue easily enough, but the reality is a lot harder. Lots of the buildings that might have to be abandoned have been homes where families have grown up, or may even have lived in for generations. We are attached to our homes, whatever and wherever home is. It is not easy to just walk away, even if we know it's necessary. Supporting whole communities to relocate because of rising seas is a tough job, and one that as a society we have not even started working on.

After Hurricane Katrina rumbled over New Orleans in 2005 – killing 1,800 people and inundating large parts of the city – there was talk of relocating the whole city, moving it upriver and away from the coast. As well as being prone to hurricane activity, the city sits at the mouth of the Mississippi River, on subsiding land, around half of which is below sea level. So, relocation made a lot of sense. But, despite the damage and the ongoing dangers, the city did not move. It was home to many – and most wanted to keep it that way. Today, New Orleans is still located where it was pre-Katrina, but sea levels have gone up a few more centimetres, and Atlantic hurricanes are getting more intense. In response, the city has invested in extensive engineering works such as levees and pumping systems to reduce the risk of inundation. It has also been partly redesigned to include wetlands and floodable parks, which accommodate water and reduce flood damage. It's possible that all this technology in the form of pumps and levees will keep the city safe – but can it last forever? Will a time come in the next century when the city really will have to pull out for good?

In New Zealand, cities like Dunedin face the same questions. Things are bound to get worse in most of our coastal cities and towns this century. Problems with stormwater drainage and coastal inundation will ramp up, and at some point a decision will have to be made to retreat – from the Kāpiti shoreline, from the Napier waterfront, from Mission Bay in Auckland, from any number of coastal settlements that have been home to many generations of people. When do we start planning to retreat? How much are we prepared to spend to stay where we are?

There are changes afoot in the policy and legislative landscape that will hopefully move us forward in our response to climate change. The Resource Management Act will be carved up into several new pieces of legislation, including something focused on climate change. In August 2022, the Ministry for the Environment/Manatū Mō Te Taiao released its first national adaptation plan, and the government has a clear agenda to help all New Zealanders deal with the consequences of climate change and best protect communities from future risks.

The responses we need to make on the ground can be very hard to deal with, and the planning timescale can be long, so the sooner we get on with thinking and talking about what we're going to do, the better. We don't have to wait for all the properties in a coastal community to be badly affected by coastal inundation before that community starts to fall apart. If just a few houses are damaged by storm waves, if part of the esplanade along the beach gets eaten away, if the bus stops going that way because it has become too risky, people are not going to want to live in that area any more. Here in my neighbourhood in Kāpiti, I imagine that once even a handful of houses on the front line of the dunes are badly affected, and even if a short section of the coast road suffers storm damage, a lot of home owners will head for the hills, literally.

There's one other thing that will eventually drive people away from oceanfront properties, and that's insurance. As *New Zealand Geographic* reported in September 2021, some research already indicates that many homes currently deemed 'low risk' – which is to say, they're estimated to flood only once a century – will nevertheless likely start losing their insurance from 2030. In a report published in December 2020, the research team at Climate Sigma, which provides scenario analysis and asset valuation on the physical risks of climate change, predicted that at least 10,000 properties across Auckland, Wellington, Christchurch and Dunedin will completely lose their insurance by 2050.

That's something it's hard to reckon with around climate change. It isn't just physical damage to coastal property, or to properties and infrastructure on flood plains, that will tip things over for most of us. Once things start to go downhill in some way, and once it starts hitting the pocket, people will vote with their feet. Neighbourhoods and even whole towns may be looking at relocation or abandonment well before the waves or the floods take over.

*

Human-driven climate change is a clear driver of increasing wildfire risk and devastation, especially in already fire-prone areas, but land use is another factor that comes into play to complicate the story.

Over the globe as a whole, the occurrence of wildfires is expected to increase by about a quarter by the middle of this century, with the greatest increases in regions that are already fire-prone. But, curiously, the total area that's burned worldwide has actually been decreasing over the last few decades – it's just that the fires that do arise have been more intense and more widespread, where they happen.

Over time, as human industry has expanded worldwide, we've cleared huge areas of land for the sake of agriculture, with forests being converted to grassland. In other places, areas that were once just bush have been cleared and replaced with housing. Where forest remains, especially for commercial purposes, practices designed to suppress fires have become increasingly common, such as grazing to keep down undergrowth, thinning tree stocks near buildings and infrastructure, improving water supplies and their delivery mechanisms, and so on.

The thing is that all of the above can swing the risk of wildfire one way or the other. Take forestry management, for example. Sometimes, the same practices implemented to reduce fire risk can actually end up increasing it, because they interfere with the natural tendency for vegetation to burn from time to time. For this reason, management practices that are more aligned with the natural cycle of fires, such as controlled and regular burn-offs, will do a better long-term job of containing fire size and intensity. Even so, they won't stop the fires – and the warming and drying brought about by climate change still increases the danger.

When it comes to housing, humanity's increasing population is forcing towns and cities to expand into areas that were once just bush. In some places, such as Canberra in Australia and Paradise in California, there has been an increase in homes built in 'the bush' – in areas that were formerly uninhabited or considered uninhabitable. Building houses in amongst trees of course raises the risk of damage in a forest fire, as was found in devastating fashion in the 2009 bushfires that burned across the Australian state of Victoria, and resulted in 173 fatalities. A portion of the death toll is understood to be related to increased population and housing on the fringes of Melbourne, in bush country. In some areas at least, we have collectively put our communities in increased danger, without really realising. Or without taking the danger seriously, perhaps – the old story of thinking it'll happen to someone else, somewhere else.

When it comes to fire, the best response is really the same as that to water: retreat. Identifying very fire-prone areas, then moving communities away is probably the best shot we have at reducing our exposure and vulnerability as fire danger increases. But, as in other areas, that is more easily said than done. Often, people live in forested environments and in fire-prone areas because these areas are also beautiful, offering access to a wide range of options, from hiking and mountain-biking to boating and water-skiing. Properties in such areas are desirable, and command high prices, just as beachfront properties do. So far, we have tended to increase the danger by allowing more development in high fire-danger areas. Future-proofing our communities will require a clear-eyed assessment of risks and how they are changing, and the willingness to change or to move if the situation or the climate requires it.

*

There's a philosophical component to our response to climate change that has to do with getting our heads around the scale of what's happening, and how it fits into the Earth's own immense timeline. We can understand what's happening in front of our faces, and we can rationalise the history of the world's climate, but bringing the two together – along with everything else that's at play – requires a kind of mental gymnastics that eludes even the smartest brains.

The Long Now Foundation is a non-profit that was, to use their words, 'established in 1996 to foster long-term thinking'. Their work aims to encourage people to imagine things on the timescale of civilisation – which is to say, over the course of the next and last 10,000 years. It's this timespan that they refer to as the 'long now'. One of their ambitious projects, The Clock, is being built inside a mountain in Texas and is designed to keep time for the next ten millennia – a feat not just of comprehension, but of mechanical engineering. Meanwhile, the Rosetta Project has brought together international language specialists and native speakers to build a publicly accessible digital library of human languages. The foundation's first foray into long-term archiving, the Rosetta Project aims to address the issue of digital obsolescence and come up with more creative and long-lasting ways to store knowledge. The project's first prototype, the Rosetta Disc, is small enough to fit in the palm of your hand, but contains over 13,000 pages of information on over 1,500 human languages – all the information microscopically etched onto the disc's surface.

In the short term, the foundation also runs The Interval, a bar in San Francisco that resides in a museum and library designed for contemplation and thoughtful conversation, and specialises in heritage cocktails. I had the privilege of going to a meeting there a few years ago, and I can confirm that it is indeed a nice place to go for a drink and to think about the future of humanity (or about tomorrow's conference talk, for that matter).

I find it heartening to see initiatives like The Long Now putting a concerted effort into getting us to lift our eyes from the ground in front of us, encouraging us to peer out to the horizon and far beyond. Who knows what the fate of humanity will be in 10,000 years? Whatever it is, I like the idea that The Clock will still be ticking.

In the meantime, I hope the world will face up to the changes happening in the climate in a way that's not just reactionary. I hope we will front-foot our response. I hope we'll see climate change for what it is, accept what can't be undone and do what we can to stop further damage, by taking action now and by planning proactively for a future in a warmer world. I hope that all of us will embrace the aim of The Long Now: 'We hope to help each other be good ancestors. We hope to preserve possibilities for the future.'

*

Here in New Zealand, a positive response in recent years happens to also be one I've had the privilege of being personally involved with. In 2019, the Labour Government established He Pou a Rangi – The Climate Change Commission, an independent Crown authority tasked with providing evidence-based advice to guide the country to adapt in ways that help address the problems of climate change.

The commission was established as part of the Climate Change Response (Zero Carbon) Amendment Act, which passed in 2019 with multi-party support, and made our domestic emissions reduction targets more ambitious and structured. 'Some issues are too big for politics, and the biggest of all is the climate crisis we face,' Climate Change Minister James Shaw said at the time. 'Our decision to create the Climate Change Commission was about protecting climate policy from political mood swings, meaning every future government can stay focused on the job at hand: to help solve climate change and make sure our communities are cleaner and healthier.'

I am one of eight commissioners, overseeing dozens of commission staff, and it's our collective job to figure out how to meet the Act's requirements in a way that balances the needs of the environment, the economy and society. No big deal. It's only the country's future that's on the line. The Board of Commissioners is chaired by former Reserve Bank Chairperson Dr Rod Carr, with Ngāi Tahu Kaiwhakahaere Lisa Tumahai as Deputy Chair.

An experienced businessman with a pragmatic and action-focused approach, Rod was the perfect choice for Chairperson – but he was actually in the process of transitioning to a new phase when he was offered the role. 'I was supposed to be retiring!' he says – but, by 'retiring', he means still working, just with a governance focus, so he didn't exactly have an empty plate. But then he mentioned He Pou a Rangi at the dinner table, and his 20-something-year-old son looked at him and said, 'You should just quit all the other jobs and do the climate change one. It's the only one that really matters. It's the only one that's going to make a difference.' So Rod took the climate change job. (And kept most of the other roles too. That's Rod – the consummate doer.)

Being on the commission can be a bit of a juggling act, where we try to balance what the science is telling us – to take decisive action yesterday – with wider forces, such as the economy and New Zealand society. As Rod says, in order for us to see real change, as quickly as possible, 'we have to convince all New Zealanders that there is a better future with lower emissions – a cleaner, greener, healthier, more sustainable, more prosperous future'.

In our first major piece of work, delivered in May 2021, the commission laid out our advice for achieving net-zero emissions by 2050. This advice was, as requested, based on what we deemed achievable with current technology, and noted that emissions cuts needed to become increasingly ambitious over time. The government heeded our advice, responding with a $2.9 billion package of initiatives including increasingly ambitious emissions reduction budgets over stages: 22.6 per cent from 2026 to 2030, and 39.1 per cent from 2031 to 2035. Between 2022 and 2025, the aim is to reduce annual emissions by 7.99 per cent against 2020 annual emissions. Funding was also allocated to projects in transport, agriculture, forestry and energy production in order to actually meet these targets. I think it's fair to say that everyone at the commission was both pleased and relieved to see so much of our advice reflected in the resulting plan.

Since then, there's been a flurry of additional work. We've issued advice on how agricultural emissions should be priced, on the Emissions Trading Scheme, and on the aforementioned national climate adaptation plan. It feels like we've made a lot of progress – but, hovering in the back of my mind at least, is the knowledge that our success depends on the extent to which we can influence the government of the day. For now, the current government is taking on board our evidence-based advice, but we have no policy levers to pull ourselves. Expert advice is only powerful when it's implemented. Rod is as aware of this as anyone. 'The challenge for us is to get through at least a couple of election cycles, where the voters determine that the work of the commission is in their interests,' he explains. 'It is in their interests not to politicise necessary climate action.'

This sentiment echoes something American environmentalist and author Bill McKibben recently noted in his newsletter, *The Crucial Years*, during the COP27 proceedings. 'Climate change shouldn't be a politically polarizing issue,' he wrote. 'Physics has assigned us a straightforward task, which is to stop burning fossil fuel, and given us a tight timeline. As a world, we should be hard at work.'

Like me, like the other commissioners, Rod is motivated by this same knowledge that time is of the essence. When he discussed taking the job of Chairperson at the dinner table, he was encouraged by his family but also confronted with a blunt question. 'My kids said to me, "Where were you for the last 30 years? You could have known. You should have known,"' he explains. 'And I was busy doing other stuff.

That's not an excuse. It's an explanation.' Right now, it's time for the other stuff to wait.

Unfortunately, the undeniable urgency may not be enough to motivate the rapid and wide-scale adaptation required of New Zealand and the world as a whole. 'I can see why my kids and their generation are exhausted and frustrated,' says Rod, 'but the fact we should have done more sooner doesn't make it easier to accelerate the change needed now.'

We had better learn how to adapt to the consequences of at least 1.5°C warming, because that's what we are stuck with so far. But the reductions we need to make in global emissions if we're going to halt warming at that point – around 8 per cent per year – are an extremely big ask. It is hard to imagine the global energy transition working faster than this. Short of shutting down whole industries and economies, and causing immense economic and social chaos, it is going be really challenging to move as quickly as we need to.

But move we must. Yes, the costs and the damage will still be substantial, and millions of people will be displaced. However, with concerted effort, the global community could likely absorb that damage, and we would have a shot at averting worse.

IV

WHERE TO FROM HERE?

13

What if?

'The window of possible climate futures is narrowing,' wrote US climate journalist David Wallace-Wells in *The New York Times Magazine* in October 2022. 'As a result, we are getting a clearer sense of what's to come: a new world, full of disruption but also billions of people, well past climate normal and yet mercifully short of true climate apocalypse.' And, while it's certainly true that the window is narrowing, I'm not so sure we've yet dodged the worst. All I can say is that I hope he's right about us being mercifully short of climate apocalypse …

We all want to know what the future holds, but crystal-ball gazing is unreliable at best. When it comes to climate change, we know what's happened so far – that we have likely locked in 1.5°C of warming – but beyond that, the best we can do is to run through a series of what-if scenarios. What if we stop all emissions tomorrow? What if we do nothing at all? What if we opt for something somewhere in between?

Well, let's start with the blue-sky option, the absolute best-case scenario: stopping all emissions tomorrow (or, even better, today). If we did that, climate change would stop shortly afterwards, with perhaps a tenth or so of a degree of warming beyond where we are now.

Comparatively, if emissions were reduced rapidly – if we got to zero worldwide, say, no later than 2050 – global warming would stop at around 1.5°C or a little more. That's the most ambitious end of the 2015 Paris Agreement range.

But, if we make no effort whatsoever to reduce emissions – and, instead, carry on merrily burning fossil fuels the way we have for the last few years – the globe will warm by about 3°C this century.

And, if the use of fossil fuels keeps *increasing* at the same rate it has the last 40 years? Then, global warming could reach 4°C or more by the end of this century. Furthermore, it could go several degrees beyond that in the next century or two. It is towards that end of the spectrum, when we start reaching 3°C or more of warming, that catastrophe surely lies.

The globe has already warmed a degree or so above pre-industrial levels. We are already seeing more intense forest fires, record heatwaves, unprecedented floods and droughts. The more warming the globe experiences, the more the climate moves outside what human society has become used to, and the more damage and risk there is for all of us. Even 1.5°C of warming – our current best-case scenario – is no guarantee of safety. There will be even more intense extreme events and perhaps half a metre of sea-level rise. However, that level of change would probably be manageable, for most countries. Probably.

*

Let's take a moment with that worst-case scenario of 3°C warming or more. That level of warming would pull the environmental rug out from under all of us, and would turn our world upside down.

The occurrence of extreme high temperatures, of droughts and deluges, would cause crop failures all over the world, knocking out global food security. Fires and floods would cause massive economic damage and put millions, if not billions, of lives at risk. The spread of what were once solely tropical diseases would put millions more lives at risk. Heat stress would also worsen respiratory illnesses, heart disease, food spoilage and food-borne disease. The mental-health impacts of an increasingly extreme climate are hard to judge, but could also be devastating. The stress brought about by the Covid-19 global pandemic has been challenging enough. Watching the world and its people suffer under 3°C or more of global warming would be almost unimaginable.

Combine all that with a certain future of many metres of sea-level rise and the displacement of a billion people, and we have a recipe for mass migration, the breakdown of the rule of law, regional warfare over food and resources, the collapse of the global economy and the collapse of civilisation.

That is a seriously frightening prospect.

Global collapse, or even major damage to global food security and the global economy, might sound far-fetched, but these are the kinds of possibilities that IPCC Assessment Reports have been signalling for at least two decades. Rather quaintly, the IPCC refer to these things as 'Reasons For Concern', which is putting it mildly, to say the least.

What about if we pulled out all the stops and burnt all the known fossil-fuel reserves as fast as possible? In that admittedly (and thankfully) unlikely scenario, the world could be 4°C warmer by the end of this century. That's one human lifetime away. On that trajectory, things would be getting difficult in just 20 or 30 years. By then, temperatures would have gone up another degree or more and extremes beyond anything we've seen in the past would be commonplace.

The West Antarctic Ice Sheet would have begun its unstoppable slide into the ocean, while the Greenland ice sheet would be thinning rapidly. Arctic sea ice would disappear, in summer at least, helping to warm the northern polar region by 7 or 8°C above pre-industrial levels. The Arctic permafrost would thaw out rapidly, releasing yet more carbon dioxide and methane into the air and only adding to the overall problem.

Other tipping points would likely come into play, too. The Amazon rainforest is estimated to be close to a point where it could die back and the region could turn into a savannah much like the grasslands in parts of Africa. If that happened, it would profoundly change the climate of a large part of South America, and would affect the circulation of the atmosphere across the tropics. It would also worsen the carbon dioxide problem, as rainforest absorbs much more carbon dioxide than grassland does.

How would New Zealand fare in such a future?

Not as badly as some. In the short-term. Maybe. Even so, it would still be extremely difficult for us.

The days we currently call 'hot' would become the norm for most of the warm half of each year. Winters would be much milder, with no frosts and a drastic reduction of cold nights. Drought would be commonplace in the east and the north of the country, punctuated by very heavy rainfalls when a storm came through, alternately baking and soaking the ground. These changes would put huge stress on our agricultural economy, and on water availability in eastern regions.

Changes to land use would be required, involving a move to dryland-style farming in eastern regions and in the Far North. New Zealand would be suitable for many more subtropical crops than are currently farmed here, but other crops may struggle or would have to be moved much further south, especially pip and stone fruit, kiwifruit and grapes.

A wide range of new pests and diseases would be able to thrive here, raising risks for public health as well as for agriculture, forestry and native ecosystems. A tripling of the period of extreme fire danger in eastern regions would pose further risks to both managed and native forests. The loss of snow and ice in our mountains would drastically change flows in major rivers and would change the look of the land, especially in the South Island. It is hard to say how many native species would be pushed to extinction, or near to it, but many would be affected.

Changes in other countries would also have an impact here. As a country, New Zealand relies heavily on its international connections, for travel and trade and to define its place in the world. If other countries are suffering more than we are, if other economies are badly affected, we will be badly affected too. A large part of our economy stands or falls with the global economy. There do not need to be crippling extreme events here before we could take major financial hits through the breakdown of international trade.

One of the factors that can only make whatever future we face worse is that not everyone will experience the same consequences at the same time. The worst-affected places would likely be some of the countries least able to afford to deal with the consequences, and also those who do the least emitting of greenhouse gases. Many tropical countries would bear the brunt of food shortages and crop failures, and water-availability problems. For instance, only half a metre of sea-level rise would displace around 15 million people in Bangladesh, while many Pacific Island countries would struggle to maintain food security and even habitable land. Inequalities are almost bound to get worse in a world where poorer countries are suffering the most. How long it would take for this to lead to regional or international conflict is anyone's guess, but it is something that is on the radar of many of the world's defence forces.

If sea levels and extreme events are causing devastation elsewhere, and hundreds of thousands – maybe even millions – of people are displaced, they could well be heading for our shores. If that happened, we would be in an unprecedented situation. What is our obligation to those from other countries who can no longer live in their homelands? Do we have one? It seems to me as though the choice would be to either open our arms or seal the borders.

This brings us back to the idea of New Zealand as a potential safe haven from climate change. While that could be seen to be true, for a while at least, it's not necessarily all good news. In fact, it is likely to go from being a blessing to a curse. If only a handful of well-off citizens of other countries choose to look for residence here, that may be considered acceptable. If hundreds of thousands or even millions of people – refugees and the wealthy alike – try to enter the country, regardless of legal requirements, that would be a very different situation. How does a country like ours prepare for something like that?

*

Writing about what such a future would be like is, it seems to me, bound to fall short of the reality, no matter how lurid I make the prose. If societies around the world are suffering huge hammer blows from crop failures, flooding, inundation, if the global economy is going belly-up, if boats full of refugees are massing in the Hauraki Gulf, I think we would all struggle to cope with it. The disappearance of the Fox Glacier would be the least of our worries. It has been hard enough the past few years, watching Covid-19 sweep across the world, infecting tens of millions. Imagine seeing news of the devastation of Shanghai, London, New York, even Auckland. How would it feel to know the wonderful modern society we have built up over centuries is now in grave danger and may be over for good?

It may sound as though I am saying, 'We're all going to die!' but not really. As grim as a future of unchecked global warming sounds, even that future would not mean the extinction of the human species. At least I don't think it would. Humans are pretty adaptable – I am sure there would be pockets of survivability, even in a lawless and hungry future. There's no way that all of us would be here in such a future though. Going from today to a world that's 4°C warmer would likely mean the death of billions of people, and the decimation of most ecosystems, on land and in the oceans. Surviving that might actually be worse than not surviving it, for those left behind.

At this juncture, I do have a sliver of good news to temper the hellfire visions: it is starting to seem as though the efforts made in recent years by climate activists, experts and policy-makers are having some effect. Where, just a few years ago, these very worst-case scenarios didn't seem so improbable, they're now looking fractionally less likely. As Wallace-Wells notes in the aforementioned *New York Times Magazine* piece, 'thanks to astonishing declines in the price of renewables, a truly global political mobilization, a clearer picture of the energy future and serious policy focus from world leaders, we have cut expected warming almost in half in just five years.' While that quote makes it sound like we're already there – and we're certainly not – it does remind us that there's reason to hope. And, most importantly, that there's still reason to act.

The reality is that we are gambling with everything we know, everything we hold dear. We are all a part of the climate system. Every mouthful of food we eat, every glass of water we drink, every breath we take – it all comes from the climate, from the environment. As a species, we are adapted to the range of climates the Earth has experienced over the last few million years. Specifically, our societies are adapted to the climate of the last few thousand years. If we change that climate much outside the range of what we know, everything around us is put at risk.

Wars, starvation, billions of deaths, unstoppable sea-level rise: that's the worst-case future we need to avoid. And avoid it we can. There is absolutely no need to go there. We are in control of this problem, and all we have to do is to stop putting greenhouse gases – especially carbon dioxide – into the air. That's all we have to do.

The catch, of course, is that modern society runs on fossil fuels, and has done for a century and more. Burning oil and coal pervades virtually every facet of our lives, and turning off the tap is a very big job. Indeed, it's the biggest job our species has ever faced. Not everyone is so sure about the need to take urgent action, even now. There's a lot of money invested in keeping going with business as usual. A lot of people believe we can carry on as we have been for a while yet, and think about taking action later, perhaps when the consequences are really staring us in the face. The trouble is, that thinking is completely wrong.

Given the stakes we're playing for, we absolutely need to rise to the challenge. There is really no choice, in my opinion. The sooner we stop emitting carbon dioxide, the sooner we stop the warming. The less warming, the less damage, the less death, the less grief. Stopping at 1.5°C would be great, and pretty manageable, though it is looking very unlikely now. Stopping at any value between that and 2°C – be it

1.6°C, or 1.7°C, or 1.8°C – would be vastly better than letting warming climb beyond that. Even stopping at 2°C would be a great deal better than missing it and letting warming climb towards 3°C or more.

When everything is at stake, there should be no limit to the speed and power of our response.

14

Global action

The most recent UN Emissions Gap Report, released in October 2022, was ominously titled 'The Closing Window'. It contained a number of alarming findings, most notably that a lack of progress on the part of the international community has the world on track for temperature rise far above the 2015 Paris Agreement goal of below 2°C, preferably 1.5°C. There is, in fact, the report noted, 'no credible pathway to 1.5°C in place'. The key message was abundantly clear: 'Only an urgent system-wide transformation can avoid climate disaster.'

And, just in case there was any doubt in anyone's mind, United Nations Secretary-general António Guterres put it even more bluntly when he spoke to world leaders at the opening of the COP27 climate summit in Egypt days later. 'We are in the fight of our lives and we are losing,' he said. 'Our planet is fast approaching tipping points that will make climate chaos irreversible. We are on a highway to climate hell with our foot on the accelerator.'

It also seems as though fossil-fuel producers only want to keep their foot down, continuing to use their power and girth to over-ride or stymie international attempts at taking meaningful action. One of the world's major oil producers, the United Arab Emirates, is set to host COP28 in 2023, and at COP27 in Egypt had the biggest delegation, with over a thousand people – that's twice as many as the next biggest delegation, from Brazil. Furthermore, it was estimated that over 600 of the delegates attending COP27 were lobbyists for the fossil-fuel industry. That means there were more fossil-fuel lobbyists in attendance than there were representatives from any of the African nations. And COP27 was, supposedly, the 'Africa COP'. People who stand to benefit from the status quo are hardly going to support moving towards a new way of doing things; instead, they will obviously use their numbers to sway action in their favour. If we are really going to break away from fossil fuels, this situation has to change.

So how do we 'urgently transform our systems', then, as the UN says we must? What can be done?

The first thing to remember is that it is not too late – in fact, it will never be too late in terms of stopping warming – but if we're going to stop at a so-called manageable level of warming, the time for action is now. And, while there are certainly things that we each, as individuals, can do to reduce our personal carbon footprints, climate change is, at the end of the day, a global problem. That means any meaningful response needs to be every bit as global as it is local and personal. The international community, together, needs to take climate change seriously and commit to preparing for a warmer future.

So, with that in mind, let's start by talking big picture – and, to do that, we'll begin with population. One of the most common questions I get asked when I give public talks is, 'Does the problem of climate change really come down to overpopulation?' and my answer is, 'Yes … sort of.'

Like a lot to do with our climate, it's complicated. To put things in perspective, global population has ballooned in the last 300 years. In the period since the Industrial Revolution, when we started adding carbon dioxide to the air, our population has gone up by a factor of ten, to eight billion in 2022. In 1900, there were around 1.6 billion people on Earth – that's roughly today's population of China on its own.

I myself struggle to visualise this rate of growth. A bit like climate change, it's not something you can properly grasp just by looking out the window. I find it helps to imagine taking China's current population and spreading it around the world – that's what the population of the entire globe was only a little over a century ago, but now there's over seven times as many of us, all crammed onto the same planet, all using the Earth's resources. That's a massive change in a short space of time – just a hundred years.

The sheer number of people on its own is a lot, but then you factor in the mass of farmed animals we grow for food. These days, humans plus livestock have a total mass of about 160 million tons (about 60 for us and 100 for the livestock). Meanwhile, wild-animal populations have gone down at the same time as ours has risen, so wild mammals have a total mass of only 7 million tons. Put another way, humans plus our farmed animals make up 96 per cent of the total mass of mammals on the Earth.

Clearly, we cannot continue to increase our numbers at such a phenomenal rate forever. We live on a finite planet with finite resources. One way of measuring this is with Earth Overshoot Day, the day on which it's estimated the global population has used up a year's supply of Earth's resources. In 2022, that day was 28 July. In other words, we are close to using twice the resources the Earth can provide and regenerate, every single year. Clearly, that is not sustainable. The Earth simply cannot accommodate an ever-growing population. Where would we all go? What would we eat? Inequalities mean we're already doing an inadequate job of providing fairly for many of those of us already here. That would only get worse with more mouths to feed.

Fortunately, growth in our numbers have actually already slowed – in the last 60 years, the population growth rate has halved, to around 1 per cent. What's more, it's projected to keep declining through the rest of the century. However, our gross numbers are so immense that even 1 per cent of annual growth will still double the global population by the end of the century – that would mean there would be 15 billion of us! Even with the growth rate steadily declining, it is still expected that there will be around 11 billion people on earth by 2100.

Can the Earth even support a population of 11 billion?

Let's put it this way: if everyone on Earth had the lifestyle and standard of living of the average New Zealander, the answer is no. There just would not be the resources to support everyone. Roughly speaking, we'd need a land area of between three and four planet Earths to give everyone in the world the average Kiwi lifestyle. If the Earth is going to support 11 billion of us – or even support the 8 billion currently here in a more equitable way than at present – big changes are required.

To that end, one of the most powerful changes that those of us from developed nations can make is to reduce our emissions, as it's developed nations that do the bulk of the emitting. Reducing emissions on an international level is critical, and it's how we'll balance the books between countries. If developed nations decarbonise rapidly, and also facilitate developing nations quickly taking up renewable technologies, we will be well on the way to a more sustainable and more equitable future.

As for whether or not the globe can support 11 billion people without changing the climate drastically? That's something we will find out pretty soon. In the meantime, we have to do everything in our power to stop emissions.

*

Back in the 1980s, major oil companies like Exxon knew very well that unabated burning of fossil fuels would change the climate beyond recognition in the twenty-first century. How? They're the ones who employed leading climate scientists to develop some of the most sophisticated climate models of the day. And what did they do with the knowledge they gleaned? Did they inform policy-makers and start investing in wind turbines?

No. Quite the contrary. They buried the scientific reports produced by their own scientists and started a campaign of disinformation and deceit that carries on to this day. Their efforts have been very effective. The science around climate change – including the incriminating role of fossil fuels – has been abundantly clear for four decades, and initially there wasn't actually any debate about whether climate change was 'real' or not. People across the political spectrum simply believed the science and got straight to talking about what needed to be done. For instance, while running for the role of President of the United States back in 1988, George Bush Senior promised during a campaign stop in Michigan to use the 'White House effect' to battle the 'greenhouse effect'. Can you imagine that today?

At the same time, Bush also pledged to convene a 'global conference on the environment' during his first year in the White House – but no such conference ever ended up taking place. So what happened? Big oil, that's what. Since the late eighties, the science community's findings and warnings have been largely ignored, all because the oil industry has in the intervening years not just argued against climate change but strategically undermined both the science itself and taking action.

Money is the driver. Of course it is. The oil industry is reckoned to be the most profitable in history, so there's a huge amount of power behind this campaign to persist with business as usual. Oil companies want to keep making money, for as long as they possibly can, whatever the consequences. The only cost that matters to them is monetary; environmental and humanitarian costs simply do not register. As a damning 2022 report published by British medical journal *The Lancet* noted:

Oil and gas companies are registering record profits, while their production strategies continue to undermine people's lives and wellbeing. An analysis of the production strategies of 15 of the world's largest oil and gas companies, as of February 2022, revealed they exceed their share of emissions consistent with 1.5°C of global heating by 37% in 2030 and 103% in 2040, continuing to undermine efforts to deliver a low-carbon, healthy, liveable future.

And, although governments now talk about climate action a lot more than they once did, the global community still subsidises the fossil-fuel sector to the tune of hundreds of billions of dollars every year. It staggers me that there are businesses, and business leaders, and world leaders who are prepared to gamble with billions of lives for the sake of profit. How long will that go on?

Even now, the fossil-fuel sector will still try to allay expert qualms and defer action by claiming that we must keep burning fossil fuels 'for the good of the economy'. They overplay any supposed economic uncertainties and crippling costs associated with decarbonising, while brushing the overwhelming and verifiable costs of continuing to burn fossil fuels under the rug. In addition, and in order to further distract, they promote the idea that individual action alone is all that's needed to stem climate change. 'Use low-energy lightbulbs!' we are told. 'Fly a bit less! Just don't fiddle with The Economy.'

I own a petrol-powered car. I fly sometimes. I buy goods imported from the other side of the world. I surf the web. I contribute to increasing greenhouse gases and climate change, just by living my life every day. Sure, there are individual actions I take to reduce my personal emissions. I eat less meat than I used to. I travel on public transport when I can, and I offset my driving and flying by donating to tree-planting schemes. I even own an e-bike, which is great for commuting. The reality is, however, that none of us as individuals can solve the climate crisis, because we all live in the world as it is – and that world is built around burning fossil fuels, because that is what has served the powers that be. It's what's lined their pockets, and they've done everything they can to hold on to the status quo. Transport, trade, technology, everything – it's all drenched in greenhouse gas emissions. It is just the way the world is, and the way the global economy is today.

If one person wants to live a low-carbon lifestyle, it is possible. You can drive an electric vehicle, install solar panels, stop flying, go off-grid and reduce your carbon footprint massively. But doing so is hard work, not to mention expensive. Right now at least, being green all on your own is difficult, and in many ways a matter of privilege. For real change to happen, what we need is a shift whereby doing the low-carbon thing becomes the easiest thing to do. When being green is just 'the way things are', then we'll be well on the way to the future we all need.

To move towards a zero-carbon world, and do it in the space of a few decades, we need active planning and guidance – and that's where governments come in. A government has some powers to help shape the economy. They can use price signals to encourage businesses to be greener. They can influence what kinds of cars we can drive, how accessible public transport is, how our cities are built, how our economy operates. There are so many avenues: a price on carbon emissions, subsidies for installation of renewable-energy plants or purchase of electric vehicles, big investments in public transport, grants for research into new zero-carbon technologies, regulations around urban design, tax incentives for low-carbon farming practices ... The list goes on. There are so many ways a government can help to shift the economy and the way we live our lives in the direction of lower carbon emissions.

Aotearoa New Zealand is in a good place already, in terms of electricity production. We have one of the world's highest fractions of renewable generation, with around 80 per cent of our power coming from hydro, geothermal and wind. Decarbonising the last 20 per cent, and building new capacity to cope with our more-electrified future, is going to require an expansion of the use of wind turbines and the establishment of solar power facilities, something we have very little of so far. As of late 2022, there are several large installations planned or already being built. There is also a lot of scope for distributed generation from solar panels on house roofs, something that is becoming much more popular as prices have reduced. There are currently no government subsidy plans for solar panel installation, but that may come soon as part of the recently established national emissions reduction plan.

There's one worry with the installation of new renewable electricity production facilities: do we have enough minerals to build all the wind turbines and solar panels? Even the amount of copper wire needed is huge, let alone the availability of rarer ingredients like lithium and cobalt, which are required for the batteries we need to store the energy from these intermittent renewables. Plus, the mining of such minerals is environmentally damaging. Even so, there are many ways to make this situation better. Massively ramped-up recycling of raw materials from things we already use, such as phones, computers and batteries, would help the supply issue and cut down on the need for mining. Scientists at CSIRO, the national science agency of Australia – which is the world's largest producer of lithium – have a major research programme in this area. New battery technologies that don't require the rarer components also look promising, and are being developed in several labs. One option currently being explored here in New Zealand is to use Lake Onslow, in Central Otago, as pumped hydro storage – or, in other words, as a kind of battery storage on a massive scale.

As with a lot of work being done in the area of climate change and low-carbon energy, we already have some amazing technologies available, and there is huge potential for innovation. My take is that we have to push ahead with these developments, and invest in new research as much as possible. To stop the climate changing, we need to get to zero emissions of carbon dioxide and other greenhouse gases – and we need to stay at zero forever! So, we need a mature renewable energy sector that will last the distance. As many people have remarked, in moving away from the use of fossil fuels, we are engineering a new industrial revolution. Any country or any business at the forefront of that massive change is almost bound to do well for itself.

The idea that a government might intervene in the economy to steer things towards low-carbon production and consumption rings alarm bells for some people. 'The market should just be left to get on with what it does best!' we hear. The trouble is, most of the environmental costs of the way we produce and consume are not properly a part of the economy.

Historically, it's been free to do all the things that cause the most damage – release greenhouse gases into the air, dump plastic waste, let nitrates leach into streams and so on – so the market doesn't really recognise them. The polluter-pays principle has never been a central part of the story, especially around climate change and greenhouse gases. As a result, *we* all pay, while the polluter gets a free ride. If our economic models start to properly account for the environment, they can become our friend in terms of tackling climate change – but, until that happens, guidance from governments is absolutely necessary.

*

The fact our economic calculations fail to account for either the climate system as a whole or the costs associated with doing things that change the climate tells us a lot about how society thinks. What our economies value is what we value – and the way we assign that value is with dollars. In this system, it's not any intrinsic value that counts; it's monetary value. There is a clear difference between valuable and being valued.

'What's really important,' we collectively say through our current economic models, 'is a strong economy, high employment and economic prosperity.' Of course, it's true that having a job and being able to pay the bills is absolutely vital to all of us, but much of what we think of as 'the economy' has become a bit divorced from reality. Share prices are essentially an expectation of value, agreed on by buyers and sellers. They can be wildly different to real value, as we sometimes discover when share prices crash. Value on paper can spike and crash without any obvious effect on the real world. Multiple times a day, we can access updates about the latest moves in worldwide stock markets, and how the value of gold has changed, but there's no mainstream coverage of what atmospheric greenhouse gas concentrations or global temperatures are doing, or what extreme events are happening in real time. Imagine how seeing those things live might change our perception of what's happening!

The problem is that the things that are of real value – the things tied to living and dying, the things that underpin everything, the whole world around us – are not included in the economy at all. The environment is a sort of nice-to-have, a green box to tick, at least in terms of government policy and business planning. I have lost count of the number of times I have heard cabinet ministers say, 'We will deal with climate change soon … we just need the economy to get a bit stronger first. Then we can afford to spend money on the environment.' As if we have all the time in the world.

Increasingly, talk is turning to how we might remodel our economies to better incorporate things according to their true value. A popular idea, which has been around since at least the 1960s, is that of the circular economy, where the need for new raw materials is minuscule because resources just go around in a reuse-and-recycle loop. As part of its programme to reduce waste across New Zealand, our own Ministry for the Environment/Manatū Mō Te Taiao has pinpointed a transition to a circular economy, noting:

> Growing international research and evidence shows numerous benefits over the traditional linear economy. These include: long-term cost savings; increased local job opportunities; encouragement of technical innovation; reducing the amount of harmful waste produced; [and] reversing our impacts on climate change.
>
> When a product's component materials are reused rather than put in a landfill, not only is that material no longer waste but new raw materials are not required to be extracted.

A conceptual model that takes things a big leap further is that of doughnut economics, pioneered in 2012 by University of Oxford economist Kate Raworth. The doughnut is a visual framework for sustainable development, consisting of two concentric rings – the inner is the 'social foundation' and the outer is the 'ecological ceiling'. Between these two sets of boundaries lies a doughnut-shaped space that is both ecologically safe and socially just – in other words, the space in which humanity can thrive. As the Doughnut Economics Action Lab puts it, 'Think of [the doughnut] as a compass for human prosperity in the twenty-first century, with the aim of meeting the needs of all people within the means of the living planet.'

These models propose a way of operating that is far from how any of us have grown up, but the mere fact we can imagine these other ways of structuring our economies demonstrates that the will to change is there. Can we do it? Can we find ways to trade and to prosper without needing to constantly grow? Can we shift away from high-emissions practices that seem like the habits of a lifetime?

I really don't have precise answers to these questions, but I firmly believe we can do it. And, of course, we have to.

*

There are more than a few silver bullets that have been suggested over the years, but I've found it pays to be wary of anything that sounds too good to be true.

For many years now, the idea of blocking out sunlight has been thrown around. Known as solar radiation management, or SRM, it's basically simulating a never-ending volcanic eruption by pumping sulphates or other light-blocking compounds into the stratosphere. Doing that would certainly cool the planet, but there are a few quite considerable downsides to the idea. First, if we pursue SRM without reducing greenhouse gas emissions, it doesn't solve the issue of ocean acidification, which will only continue worsening. Second, tampering with the amount of sunlight that falls on the Earth will change global rainfall patterns, and cause problems for agriculture and food security. Third, once we started down this path, we would have to keep going with it forever. If we just kept increasing carbon dioxide levels then stopped SRM at any point, there would be a sudden increase in temperature across the globe. Climate researchers call this termination shock, and you can imagine it would not be good news.

So, to put it bluntly, SRM is not a good idea. Just the philosophy of it seems wrong. To counter the effects of one lot of rubbish we're dumping in the atmosphere, we'll dump another lot of rubbish higher up? That makes no sense at all to me. Especially when there's already a much more straightforward and sensible idea: just turn off the flow of the first lot of rubbish (the greenhouse gases).

One other notion that has been toyed with is finding

a way to remove carbon dioxide from the air. What if we scattered iron compounds over the global oceans to spur phytoplankton blooms? The plankton would soak up carbon dioxide as they grow, then take it to the ocean floor when they die. Sounds like a great idea … but turns out it doesn't work very well. Scientists at NIWA tried this in the ocean south of New Zealand several years ago, and it worked at first, but plankton growth is limited by the lack of other nutrients, not just iron, so it was impossible to achieve blooms on the scale we'd need. So what if we then spread a cocktail of nutrients in the global oceans? Unfortunately, the results would still be unlikely to result in much carbon dioxide being removed from the air. And, even if it did work, we'd have to keep it up forever, just like with SRM.

Perhaps we could speed up the geological processes that absorb carbon dioxide, such as rock weathering? Researchers are currently looking into ways that we might do this, potentially through rock grinding or by chemical means. Or maybe we can remove the carbon dioxide from the air directly through chemical processes? These and several other carbon-extraction ideas are being investigated, but they all require large amounts of energy and possibly water, which would make them hard to roll out on the scale that would be required.

The thing with carbon dioxide, as with all the greenhouse gases, is that it comprises a vanishingly small portion of our atmosphere – only 0.04 per cent. That means we would need to process just about the whole of the global atmosphere in order to remove a noticeable fraction. Putting carbon dioxide in the air is a lot easier than taking it out again. It's a bit like making a curry – the spices you add to the pot may make up only 0.04 per cent of the whole pot of curry, but they affect the flavour profoundly. It's easy to add the spice to the pot, but getting it back out again once the curry is cooked is extremely difficult, if not impossible.

The bottom line when it comes to removing carbon dioxide from the atmosphere is that we cannot bet on a technological fix. If one could be developed, that would be fantastic – a real game-changer – but there's nothing realistic on the horizon. Well, nothing that didn't already exist before we set off this whole climate change situation in the first place. There is one 'technology' that does work extremely well: planting trees. As they grow, trees soak up carbon dioxide and emit oxygen (handy for us). Restoring wetlands or using farming approaches that boost carbon storage in soils similarly play a role in absorbing carbon dioxide. These approaches can't wholly offset our emissions – we emit far too much for that – but, at the very least, they would buy us some time by slowing the increase in atmospheric carbon dioxide levels.

Time that we could put towards the one response that is 100 per cent proven to be effective: turning off the tap of greenhouse gas emissions.

15

Local action

Now that we've talked about what can be done on a global level, that brings us to New Zealand's role in all of this. The global actions mentioned in the previous chapter can also be local actions, things that we as a country can prioritise and implement. But is it worth even bothering? Do our emissions really matter, when we are so small?

To answer these questions, we need to reflect on what is happening to the climate. Every country emits greenhouse gases, and emissions have been going on for nearly 300 years. The climate is changing everywhere, and that affects all of us, in every country. We really are all in this together. It is going to take effort from all of us, in countries and communities large and small. The best thing we can all do is to cooperate, to help and support each other, to be empathetic. To love one another, basically.

Yes, some countries and some industries have contributed more than others to the problem we face today, and the communities least to blame are the ones suffering the most. But, while it's true that those who have emitted the most should be taking the lead, it's also true that we cannot wait for what should happen.

New Zealand is particularly well placed to make a true difference not just for our own good but for the good of the whole world. Our gross national emissions might pale in comparison to bigger places like the US or China, but per head of population we fare much less favourably: per capita, New Zealanders are in the top tier of greenhouse gas emitters, and our economy has done well on the back of all the fossil fuels we've collectively burned. We therefore have at least as much responsibility to act as anyone.

Taking up that responsibility is, I believe, a huge opportunity for Aotearoa. If any country can become carbon-zero, surely it is ours. We've got a small population, with a small and agile economy. We are super well-endowed with renewable resources – plenty of wind, sun, water and geothermal energy. We are smart and innovative. We have demonstrated over and over that we can be leaders on the world stage, on terrorism, on the nuclear threat, on social welfare. We can definitely lead the world on climate change as well.

We just need to be brave, and take up the challenge. Let's do this!

*

'But what we do or don't do makes very little difference to the climate system,' some will say, overlooking the fact that the most powerful thing that every single nation in the world can do right now is get on track towards a zero-carbon future.

To put things in stark perspective, in 2021, the world collectively put around 41 billion tons of carbon dioxide into the atmosphere. Of that, New Zealand contributed just over 37 million tons. If the global community wants to stop warming at around 1.5°C – the goal at the heart of the 2015 Paris Agreement – we need to make enormous cuts, reining in emissions by 45 per cent by 2030, and getting to zero emissions of carbon dioxide around 2050. If we want to keep warming under 2°C, the upper end of the Paris target, we need to cut emissions by 30 per cent by 2030, and get to zero emissions by around 2070.

Getting to zero emissions is the only way that we are going to halt global warming. That's true zero – not just a reduction. To get to zero by 2050 and get halfway there by 2030, we need to be reducing global emissions by around 7 per cent every year, starting right now. Here in New Zealand, we need to reduce our national emissions by around the same amount every year.

So how long do we have to get started? Well, we definitely don't have Bowie's five years left to cry in. We've got zero years. The longer we wait to get started, the harder it's going to get, and the more drastic the emissions cuts we'll have to make. If the world community doesn't really get started on emissions reductions until 2030, we'll be at 1.5°C warming, well on the way to 2°C warming, and we would then have the same race to reduce emissions, just less time to do it. If we're to have any chance of real success, action has to start immediately – better, it would start yesterday. As it is, we're already going to see more extreme heat, fires, floods, crop failures as this decade progresses. If we do nothing, we may have 20 or 30 years before things get really ugly.

So far, New Zealand has achieved very little in terms of overall emissions reductions. Total emissions are slightly down on a peak in 2005, but total gross emissions and net emissions (after subtracting offsets from tree-planting and other land-use changes) have changed little this century so far. Where we have made progress is in the legal and policy settings. We now have zero net carbon dioxide emissions by 2050 as a legal requirement, with 24–47 per cent reductions in biogenic methane by 2050, under the revised Climate Change Response (Zero Carbon) Act. And, as mentioned, that Act also set up the Climate Change Commission, which has already delivered a roadmap for emissions reductions through to 2035 and has advised on methane emissions pricing and Emissions Trading Scheme settings. The government has in response published its emissions reduction plan, and has drafted its first national adaptation plan.

So, we are at the starting blocks. Now, we need to see action.

In particular, I really want to see national emissions start to come down in 2023. Reductions in carbon dioxide emissions can come mostly from the transport sector, and from industry and energy production. Here's what I would love to see happening over the next few years – call it my wish list, if you like.

- **Big boosts in renewably powered public transport across the country.** That means strongly upgraded bus services in cities, and the introduction of light rail where appropriate. Combine that with big investments to breathe new life into the national rail network, moving people and freight across the country in a network even better than we had when I was a kid hanging out at Springfield Railway Station. We also need to see improvements in the inter-city bus network. Essentially, the aim should be for people to get from wherever they are to wherever they need to be almost exclusively by public and renewably powered transport.

- **Improved facilities and incentives for active transport.** For places not covered by the transport network outlined above, we need to make it easier and more appealing for people to get around by walking, cycling or similar. That means things like dedicated cycleways, clear and safe walking paths, and subsidies for the purchase of non-car vehicles like e-bikes, e-scooters and e-skateboards.

- **Comprehensive and affordable access to car-sharing schemes.** For those who need a car to get around, it needs to be easier to either borrow or share one. By providing this option, along with improving access to public transport, we'd almost do away with the need for anyone to own a car!

- **Incentives for purchasing electric vehicles and a national charging-station network.** I've deliberately put this point after the improvements to public and active transport, as the future is not about each of us swapping our petrol-powered cars for electric equivalents. We need to get out of our cars, or the congestion will continue. Cities in many countries have already shown the benefits to be gained by kicking the car habit. We just need to join the trend.

- **Big investments in solar and wind power.** When it comes to solar generation, we need to see both large plants and panels distributed across private home roofs. We also need increases in wind generation. If we set our minds to it, we should be able to turn off the Huntly power station this decade and go to a 100-per-cent-plus renewably powered grid before 2030, one that has even more generating capacity than we have now. Imagine that!

- **Better home insulation.** Part of the future energy equation is saving energy, as well as increased renewable generation. Houses that are insulated properly require far less energy to heat and cool, but New Zealand's current building standards do not meet the conditions healthy and energy-efficient homes require. To improve the standards, the legislation needs to change, and subsidies and incentives for building climate-conscious housing need to be prioritised.

- **More efficient home appliances.** Tighter standards for how much energy home appliances use goes hand in hand with saving energy.

- **Goodbye, coal boilers!** Where they are used – in schools and other facilities, and for industrial heat – coal boilers must be replaced by solar and other zero-carbon energy production.

- **More climate-conscious development.** Changes to legislation are needed to encourage increased urban density and the end of suburban sprawl. We also need to find new ways to encourage infrastructure that allows adaptation to climate change, as well as facilitating the reduction of emissions.

- **More climate-conscious land-use.** We need to change the way we use the land to work with regional climates more than we currently do. In my opinion, for instance, the dry country in the eastern South Island would be better used for growing grain than for dairy farming supported by massive irrigation schemes. Moreover, it would be great for our agriculture sector to move to supplying a more plant-based diet, raising less meat and more vegetation, grains, pulses, vegetables and so on.

That should do for a start!

I know it will take many years to realise some of the things on this wish list, so the sooner we get started, the better. Dealing with climate change is a long-term commitment but it has real urgency up front. If we can make these changes over the next decade, it will put us in a good position for the future. And it will give us things to export – our technologies, our policy and planning processes, our philosophy generally.

If we can use what we do to help and inspire other countries, even better.

*

One of the most rewarding aspects of my career over the last two decades has been contributing as a lead author to the IPCC's Assessment Reports. These massive reports involve many, many hours of work on the part of thousands of scientists from around the world, and they have become a vital tool for building the case for action now.

Assessment reports are released every six or seven years and are, by nature, very technical and thousands of pages long. Crucially, each one is accompanied by a Summary for Policymakers, a document that usually runs to no more than 30 pages or so, and distils the central messages of all those observations and scientific analyses in order for senior policy advisors the world over to understand the bare essentials. This important summary is what's used to brief politicians – those who have the power, if not always the will, to speed up the process of decarbonising the world's economies.

However, as useful as the summary has been in guiding policy decisions here in New Zealand, there was, until recently, a disconnect between the scientific advice and actual on-the-ground decision-making. Various government departments took responsibility for providing climate-related advice on everything from the Emissions Trading Scheme to public transport and fisheries, but no one body or agency was tasked with taking a big-picture view, considering the evidence and – most importantly of all – outlining pathways to a low-carbon world. The establishment of the Climate Change Commission has started to change that, and has restored some of my hope that New Zealand may actually be able to step up to the immense challenge facing us and our world.

Some might say it's not our job to get to zero, that we're doing fine. 'What can we do? Globally, we are small players.' But, being small players has never stopped us on the sports field, so why should it stop us when it comes to climate change? Just as we do in the sporting arena, we have a chance here to play a key role in leading the change. We can set the trend, just by providing leadership – thought leadership, political leadership, social leadership. Currently, there is a lot of talk about living up to the Paris Agreement limits, about becoming carbon-neutral, but no country has actually demonstrated those things in real life yet – and that can make it hard to believe in. It can be hard to believe that a zero-carbon future is a real possibility. As the old saying goes, 'We can't be what we can't see.' But, all we need is for someone to show us the way. Once that happens, I think it'll open the floodgates. Every country will want to join the parade.

If any country can become 100 per cent carbon-neutral, surely it's Aotearoa. And just imagine if we did manage it! Imagine if we did it before any other developed nation! We could even do it before 2050 … We'd win the zero-carbon world cup, and we'd also garner international attention and investment. We could become a global hub for green technology and innovation, for innovative policy, for sustainable urban design, for sustainable low-carbon agriculture. The sky is the limit.

There is no reason at all why New Zealand couldn't be the country to show the rest of the world how it is done. We have already demonstrated that kind of leadership before, with our responses to Covid-19, the Christchurch terror attacks, nuclear testing.

Why not be that brave little country yet again, at this time when the world needs inspiration more than ever before?

16

Personal action

I've said that no one person on their own can halt global warming, but our nations and our world are comprised of individuals. We are those individuals, and we can – and should – still take personal action. Perhaps one of us on our own cannot change everything, but when we come together we absolutely can. As they say, the biggest changes always start with the first step. It is never too early to take that first step, even if you do it alone.

It's also important not to let the 'global' part of climate change cow us. As Swedish climate activist Greta Thunberg famously commented after COP26, held in Glasgow at the end of 2021, 'We can no longer let the people in power decide what is politically possible. We can no longer let the people in power decide what hope is. Hope is not passive. Hope is not "blah, blah, blah". Hope is telling the truth. Hope is taking action. And hope always comes from the people.'

In that sense, each and every one of us is a little glimmer of hope. There is a lot that each of us can do, especially to demand better of the powers that be. One of the most effective ways to make a difference is to do what Greta does: speak up. Talk about climate change, and about what's really happening. Share your concerns. Make sure your family, your community, your workmates know what's going on. The more we talk about climate change, the more we all care about it, and the more likely we will do something about it.

Most importantly, talk to our elected officials. Get active politically. Governments all over the world are very good at talking the talk – the blah, blah, blah – but they need to be pushed to really walk the walk. Always. Email anyone you can think of who has access to the levers of power: your local council officials, your local MP, the Minister of Climate Change, the Prime Minister. Be vocal about your concerns. Let them know that, for you, this is a top-of-the-agenda issue.

Go on a march, if there is one happening. If not, organise one yourself! As American anthropologist Margaret

Mead said, 'Never doubt that a small group of thoughtful, committed individuals can change the world. In fact, it's the only thing that ever has.'

*

Beyond communicating and a bit of activism, we can all work to reduce our own personal carbon footprint. This kind of action is important too, even though we really need a total overhaul of our economies and systems.

If you can tell yourself that you are personally helping, even a little bit, that's a plus. On the other hand, if you take no action at all, you will tell yourself that – you'll buy in to the 'What difference can I possibly make?' myth – and I don't imagine that's great for anyone's sense of self-worth, especially if you are concerned about climate change.

My philosophy is that anything that makes a difference makes a difference. If we all reduced our personal emissions by five per cent, that would be a big step forward for the whole country. So, whenever you can, make the low-carbon choice.

I'll give you some more concrete suggestions below, but before I do I also want to add this: if you can't do these things, please don't despair. None of us is 'bad' for contributing to greenhouse gas emissions. The way our world is structured means that, most of the time, we actually can't help it. Feeling guilty or victimised really isn't helpful for any of us.

No one is a baddie and no one is a saint. None of us is emissions-free, and tackling climate change is not about lecturing or judging others. Finger-pointing is a common – and nasty – ploy by those who don't want anyone speaking out about climate change. Some say for instance that, since I drive a petrol-powered car, I can't possibly talk to others about emissions reductions, but that's nothing more than an attempt to distract from the true issue at hand.

We all have a role to play. It's our job to support one other. It's our job to do whatever we can, however we can.

*

In New Zealand, one of the most powerful ways each of us can take personal action is by being more thoughtful in our transport choices.

For decades, transport has been our country's biggest growth area for carbon dioxide emissions, especially from private cars. There are a few ways you might be able to reduce your personal transport emissions:

- You can drive less
- You can use public transport more, especially if it's renewably powered
- You might also be able to walk or bike more
- And you might be able to buy an electric car or a more efficient car.

Lots of these things have positive side-effects above and beyond reducing emissions. On a bus or a train, you can read a book or listen to a podcast or do some work or even, if you're so inclined, take a nap. Getting some active transport into the day is great for general health and fitness, and a bit of fresh air does wonders at the end of a long day.

One area of transport that gets a lot of bad press is flying, and attached to that is the idea of flight shame – causing others to feel bad about travelling in planes, ever. Of course, flying is a luxury only a small portion of the world's population can afford. It's estimated that less than 5 per cent of the world's population flies – and only 1 per cent is responsible for 50 per cent of the total emissions from flying. Perhaps the most egregious examples are the tiny and wealthy minority who fly by private jet, and in recent years the private-jet usage of a number of celebrities and performers has been closely scrutinised (and criticised) online. For the vast majority of the global population, flying is not a part of their carbon footprint at all – but, for those who do fly frequently, it is a very large component of their footprint.

While it's true that choosing to fly less, especially if you tend to fly a lot, is a good way to cut your own emissions, it's a bit of a tough subject in a country like New Zealand. We are at the bottom of the world, far away from most other countries, so in order to get anywhere we have to travel long distances. Taking an international flight can add hugely to a person's carbon footprint, but many of us have family overseas and have very good reasons to travel so far. And, when it comes to going overseas, there's not really another viable option.

Even within New Zealand, air travel may be the only practical option a lot of the time. Our passenger rail network is just an echo of what it once was, and the same goes for the intercity bus network. Driving from one end of the country to the other is very time-consuming and still pretty heavy on the emissions if you drive a petrol- or diesel-powered car. To get around the country, flying makes the most sense a lot of the time.

Fortunately, there are some positive signs on the flying front. For short-haul flights, renewably powered aircraft are starting to appear. There are also already hybrid-engine planes and a few fully electric aircraft around. So far they're small and cannot fly very far, because their batteries are so heavy, but it's reasonable to expect that such craft could work on some routes in New Zealand, where distances are short and passenger numbers modest. The aviation industry is also working on developing more sustainable fuel, in the form of a mix of biofuel and traditional avgas. Plans are modest, with an expectation of 20 per cent biofuel in the mix by 2040 – that would reduce aircraft emissions a bit, but would still be far from zero.

Fully solar planes are out there, but they are years away from being ready to take passengers. In 2018, the Airbus Zephyr flew for almost 26 days straight with no fuel at all, just solar panels. A great achievement ... but the plane is basically a glider and weighs only 75 kilograms all up, less than what one typical airline passenger weighs. Right now, it seems unlikely that such an aircraft design could be scaled up to fly 200 people to Australia. With lighter-weight batteries and more efficient solar panels, the solar-plane market may grow, but it'll be many years before anyone boards at Auckland Airport.

So, it looks as though aviation will be a high-emissions sector for many years to come. What's more, the number of international travellers is only expected to grow. Right now, aviation contributes only a few per cent to total global emissions, but its emissions will become more and more pronounced as other sectors decarbonise. The fact remains that, just as all countries must get to zero carbon dioxide emissions, so to must all businesses and all sectors of the economy.

*

Another area we hear a lot about in terms of our carbon footprint is what we eat. Food production globally is responsible for about a quarter of total emissions, and there is room for lots of improvement.

You probably already know that red-meat production has the highest carbon footprint of any form of food. Beef is the highest, followed by sheep meat. White meats such as pork and chicken are much lower carbon – only about one tenth of the emissions from beef. Even a dairy product like cheese has less than half the carbon emissions of beef. Vegetables, grains, legumes and nuts are some of the lowest-carbon options. A kilo of beef has over 100 times the emissions that come from producing a kilo of potatoes or carrots.

The emissions that come from transporting food usually work out to less than 10 per cent of the total emissions from the whole supply chain, from growing to marketing and sale in shops. So, what we eat is much more important when it comes to carbon emissions than where our food comes from. There are many reasons you might choose to buy local, but your carbon footprint isn't one of the most important.

Becoming vegetarian or reducing your meat consumption – especially red meat – can reduce your overall carbon footprint by a huge amount. And doing so is a win-win: what's good for the climate is also good for your health.

*

Another way you might take personal action is by painting your roof.

You'll already know that white surface reflects a lot of sunlight, while a dark surface absorbs most of it. Painting the roofs of houses white has long been promoted as a way to help cool the Earth – in California, flat-roofed buildings are required to have white roofs, and some cities across the US and elsewhere are promoting a white-roof policy.

The results can be mixed. The idea is that reflected sunlight won't warm the Earth, which is correct. But, if you live in a polluted area, the reflected sunlight might be absorbed anyway by air pollution particles as the light heads back towards space.

A white roof will keep a building cooler, reducing energy demand for air-conditioning. On the other hand, in some climates, it may make buildings overly cool in winter, increasing the need for winter heating. It's also important to remember that all the world's roofs together cover only a tiny fraction of the Earth's surface, so the cooling effect would be very small – less that one tenth of 1°C.

Overall, here in New Zealand, where air pollution is not such a problem and winters are not especially cold, painting your roof could be a good idea. It's certainly not a terrible one. Every little bit helps.

*

Some younger people are choosing not to have children, sometimes out of a desire not to contribute to the growing population and our rampant consumption of the Earth's resources. This decision can also stem from the fact that some don't want to bring up a family in a world in crisis over unchecked climate change. It is a tricky topic, as our effect on the environment is not just about numbers of people; it is, importantly, also about levels of consumption. As I've noted, the average New Zealander or American consumes vastly more resources and emits way more greenhouse gases than the average citizen of many African countries does.

It has been estimated that the single biggest thing any of us can do to reduce our carbon footprint is to have one less child, for all the future emissions that are saved. But, whether to have a family or not is a very big and a very personal decision for anyone to make. If you do have children, they can be a reason to do better, to work harder to fix the environmental problems that all of us – including future generations – are facing. I have one child, and I want to see the best possible world for him through the rest of this century.

At the end of the day, each of us must do what feels right. Each of us must do what we feel we can.

Conclusion

Is it possible?

Back in 2013, the IPCC released its Fifth Assessment Report. I'd been involved with writing one of the chapters, so I did a bit of media after the report was made public.

'Can you run me through the key points?' one journalist asked when he called me for comment.

'Sure,' I replied. 'The report documents unprecedented warming, extreme temperatures, extreme rainfalls and sea-level rise, and the urgent need to reduce emissions of greenhouse gases, among other things.'

'Oh. So, same old, same old?' He sounded unimpressed. 'None of that sounds very newsworthy.'

I was staggered. 'It means the future of humanity is in danger. How is that *not* newsworthy?' I raised my voice a bit. 'This is the most important story ever! It should be front-page news in every newspaper, every day, everywhere, until we get a solution.'

Unfortunately, my impassioned retort didn't make that much of an impression, as the newspaper only ran a small piece.

And now, a decade on, I have to admit to feeling a bit 'same old, same old', too – I look at what the world has done in response to the biggest threat we've ever faced … and it's all very 'same old, same old'. The conversation has shifted in those ten years, but for the most part we're still carrying on with business as usual. There are a lot of issues we're more aware of now than we were a decade ago. Every one of them is important. But, the catch is this: climate change is the only one among them that will upend all our lives. And it will make every other problem worse in the process.

There's an insidious slow-creep aspect to climate change. There's a fire here, a heatwave there … then nothing obvious for months. Years pass like that, with only a few events crossing our radars – and, a lot of the time, they happen 'over there', to someone else, somewhere else. In the meantime, carbon dioxide continues to accrue in our atmosphere, and with every bit more we release, we lock in ever more change, change that will last for centuries. It is a gradual slide into ruin. If we wait for the impacts of climate change to actually be in our collective faces, it will be too late. We'll already be well on the way to a catastrophic future of 3°C or more of warming, and the end of everything we know.

The past couple of years have already been difficult ones for our world. The global Covid-19 pandemic has brought suffering, death and upheaval on a scale that many of us alive in countries like Aotearoa have never before experienced. But the pandemic has also taught us a lot about humanity's capacity to respond – or not – when faced with catastrophe.

Things started slowly with Covid, too. At first, it was a problem in one city in China. Then, maybe in a few countries far away. Then, on 28 February 2020, a case was confirmed in New Zealand – but it was just one case, no big deal. That very same day, I flew from Wellington to Nelson for a family wedding, and not for a moment did I consider that both flying and attending weddings would be completely out of the question within weeks. When the New Zealand Government put the country into lockdown less than a month later, it was unprecedented. To some, it seemed like an over-reaction.

But, by taking brave and decisive action, our government took care of our national community and saved many lives in the process – one study conducted a year later by researchers at Te Pūnaha Matatini suggested that, if we hadn't gone into lockdown so swiftly and comprehensively, Covid-19 would likely have infected a third of the population and killed more than 30,000 people by the end of 2020.

As a country, we remained mostly Covid-free until a safe vaccine was successfully developed, and could then be systematically rolled out to as many people as possible – again, thanks to bold leadership on behalf of our government. Not everyone agreed, but in this instance the most important thing was protecting as many of us as possible. Unfortunately, when faced with a crisis, not everyone's personal preferences can be accounted for; we must band together for the good of our communities as a whole. Climate change is, in that regard, no different.

The vaccines in themselves, actually, are yet another example of how quickly and effectively humans can respond to threats. In the space of months, the world's leading medical research labs developed not one but a handful of viable vaccines – a process that would ordinarily take at least a couple of years, but in this case took mere months. What got things moving so much faster? All of the world's experts threw all of their brainpower behind it at the same time. Funding was readily available for research – no need to waste time filling in proposal forms or bidding for resources. Now imagine if we did the same thing in response to climate change. How quickly could we have real, workable solutions on the table for reducing emissions and getting things moving forward? Quickly enough, I'd hope, to maybe make up for some of the lost time of the last three decades.

As well as showing us just what humanity is capable of when faced with a crisis, Covid-19 also gave us a window on what a world free of emissions might be like – and it was wonderful. As a result of international lockdowns and travel restrictions, freight and travel dropped off a cliff. No one was really flying anywhere, and no one was driving much, if at all. And so, in 2020, carbon dioxide emissions decreased by around 7 per cent and air quality around the world improved noticeably. People in New Delhi could see the Himalaya for the first time in many years. Even here in New Zealand, during the lockdown in March and April 2020, air quality in New Zealand's major cities improved by 80 or 90 per cent. And, in some of the world's most polluted cities and industrial areas, it took only a matter of days for the skies to clear.

If we can decarbonise our energy production, that's a mere taste of what the skies will be like. All round the globe, the skies will clear, without any need for lockdowns. And, climate change aside, the benefits for public health will be immense. Just those few weeks of lockdown in 2020 are estimated to have saved nearly a million premature deaths and to have avoided well over a million cases of asthma among the world's children.

Can we lock in those kind of gains in air quality again? Can we reduce greenhouse gas emissions on that scale and more?

Well, we obviously don't want a global pandemic to be the thing that motivates us to do so. Climate change on its own should be more than enough. The lockdowns showed us what could be, and the Covid response – both on an international and a national level – showed us what we can do when we worth together. So can we reduce emissions without the lockdowns and the pandemic? I believe we can.

There are already positive signs. The cost of renewable energy has dropped so fast in the last decade that it is now cheaper than coal, cheaper than nuclear, and should be the preferred option for energy companies everywhere. In April 2020, Britain went for two months generating all its electricity without using any coal at all; only a decade earlier, 40 per cent of the country's electricity had come from coal burning. And, on Boxing Day 2020, Britain had, for the first time, over half its electricity supply delivered by wind turbines. I can remember not so many years ago hearing an energy pundit say on the radio that wind power would never top 10 per cent of electricity supply – well, that's been proven false. What is possible is changing fast, and that's real cause for optimism.

Plummeting costs plus the imperative from the Paris Agreement for every country to take action on reducing emissions have led to exponential growth in renewable electricity production, especially from solar panels. According to the International Energy Agency, 'By 2026, global renewable electricity capacity is forecast to rise more than 60% from 2020 levels to over 4,800 GW – equivalent to the current total global power capacity of fossil fuels and nuclear combined.' There's still a long way to go, as fossil fuels still supply around 80 per cent of global electricity but, if the growth in renewables continues the way it has lately, burning fossil fuels will become a thing of the past.

I also suspect that once global emissions do start to drop, once we have proved to ourselves that it is actually possible to reverse a century of ever-growing pollution, we'll trigger a number of feedbacks. We'll pass a social and economic and industrial tipping point. Part of the problem now is that, for all the talk, emissions have not started to decrease. It's easy to think, 'It's all too hard. Maybe we should give up? What's the point in trying?' But all we need to be inspired is to see a sliver of actual progress. If we know we are at the start of the right path, progress will accelerate. That's my hope.

In terms of climate change, the world is where New Zealand was with Covid-19 in early March 2020. There were warnings, but not much to see actually happening – yet. With climate change, the world has months and years to act, instead of days or weeks, but we still have to start immediately. The urgency is the same, but the task is vastly bigger. Instead of locking down a relatively small population and a relatively small economy for a matter of weeks, our collective task is to transform the whole of the global economy and steer it on a path to zero greenhouse gas emissions. To convert all the energy production the world needs to renewable sources. To lock that in for all of future history. Forever.

*

So, is it possible? Can we really halt climate change? Is a carbon-zero future actually on the cards?

My answer is a cautious, but hopeful, yes. It is possible, but it is not going to be easy.

Transitioning to a world that doesn't rely on fossil fuels is going to call for more than just promises. It's going to take committed and focused action, and real change, especially from those who have benefitted most from the status quo. If New Zealand and other countries around the globe are actually going to get to a zero-carbon future, we need change from both the top down and the bottom up. If government sets the tone and the direction, we can all play a part in making that future happen. We saw this happen with the Covid-19 lockdowns in 2020. The government took urgent and decisive action, and the whole population played their part. It worked. It can happen again with climate change.

There are no free lunches here, however. Building all the solar panels, wind turbines, batteries and other hardware we're going to need will take time, energy and resources. We'll require vast amounts of copper and other minerals like lithium. Much of that will have to come from mining and, initially at least, that extraction process will need to be powered by fossil fuels. However, as the renewable transformation takes hold, more and more of the necessary energy can come from renewables themselves. Likewise, a lot of the materials we need can also come from recycling existing electronics – smartphones, tablets, computers and so on.

We're also going to have to get in the habit of using less. One of the most powerful ways we can do more with less is to design more efficient appliances and build better-insulated homes that require less energy to run in the first place. As well as shifting society away from high-emissions practices that seem like the habits of a lifetime, we're also going to need to transition to a truly circular economy, where nothing new is created and everything is just reused and repurposed in an ongoing loop.

And one last thing: to really get things moving, we need to dethrone big oil. The petro-states have controlled the conversation on climate change for far too long. They have spread disinformation. They have claimed that individuals – not big business or governments – are the ones who are responsible for responding to climate change. They have claimed just about anything, in fact, to slow down action. The fossil-fuel non-proliferation treaty proposed by Tuvalu and backed by a wide range of parties is what is needed to phase out the use of oil, coal and natural gas – but it will be a tough fight.

So, you see what I mean: it's not going to be easy.

But I firmly believe we *can* do it. And, of course, we have to.

What keeps me optimistic is knowing how endlessly inventive humans are. Just look at the world around us. Now just imagine what we could achieve if the world economy invested trillions in greening our societies, if the collective brains of the world's research labs and universities were focused on making the difference we need to get to zero carbon.

I also know I'm not the only one who believes we can do it. As we see more devastation from weather and climate extremes, as the IPCC reports become ever more alarming, people are getting active. More and more of us are concerned. More and more of us are doing something about it. There's the School Strike for Climate movement, Extinction Rebellion, Generation Zero, 350.org – so many communities, both here in New Zealand and abroad, are demanding action. (If you want to know more about these organisations, or what you can do personally, turn to the Resources list on page 305.)

Local communities and whole countries alike are declaring a climate emergency and, while those declarations don't solve the problem, they are not meaningless. They set the tone, the direction of travel, for everyone to see. Ten years ago I could not have imagined the New Zealand Government declaring a state of climate emergency, but that is what happened in 2020. And as of 2019 we have a Climate Change Commission and a law that says we must get to net-zero carbon dioxide emissions by 2050. Now we need actions to follow the rhetoric, and that's where we, as individuals, step in. It is the job of citizens everywhere to hold their governments to account. That is one of the most powerful ways that we can each make a difference.

*

Picture this: a world where all our electricity comes from the sun, wind, water and geothermal energy; where our cities are contained and high density, and comprehensively served by an extensive network of bike paths and walkways; where, no matter where you are, reliable and frequent public transport will take you door to door at any time; where all our cars are electric; where everything is recycled; where low-emissions farming produces high-value food crops at the same time as building up soil carbon; where short-haul flights are via electric or biofuel-powered planes; where more raw materials come from recycling than from mining; where no one burns fossil fuels any more because it's become both socially unacceptable and prohibitively expensive; where 'climate change' is no longer something we talk about, because it isn't happening any more.

Imagine that.

A world where climate change isn't happening.

That would be a whole new world for all of us, but it would be especially revolutionary for those of us under 30. Whenever I talk with school students, they inevitably comment on how 'climate change has always been there' for them. I would love for that to change.

Whenever I picture this carbon-zero future, I have to pinch myself. It's so hopeful, and so within our grasp. Why would we not want this? Why are we having to talk about this at all?

There are so many questions that come to mind – questions I don't have any good answers for, because I don't think there are good answers.

Why didn't the world start taking comprehensive action to reduce emissions and green the future decades ago?

Why are we so slow to start, given we know literally everything is at stake?

Isn't this the perfect opportunity for heroism? For bravery? When Covid-19 became a clear and present danger, governments all over the world took rapid action, backed up by community action. Likewise, in times of war, countries have re-tooled their economies virtually overnight to begin manufacturing weapons of war.

What's the difference with climate change?

Of course, the difference is multi-faceted: the illusion of having time, the power of profit, the fear of change all conspire to slow things down. We know the oil industry has played a big role in keeping the brakes on – putting profit ahead of the future of humanity – but, beyond that, I think a more general reason for our collective slowness to act is the fear of change. Most of us are conservative by nature – we'd rather stick with the status quo than make changes in our lives. Change takes effort. Change is hard. Change is scary. We'll defer it for as long as we can. 'I can deal with climate change another day,' we tell ourselves. 'I have other, more pressing problems to deal with right now.' That's what we think, until it's our home being burnt to the ground in a wildfire or washed away in an extreme flood, until it's our community dealing with a crippling drought.

We are all 'too busy'. I feel that in my own life: my days have definitely become faster and busier, filled with more stuff and more tasks than they once were. There are so many demands on our time, so many distractions – who has time to add one more worry to the list? I get it.

But here's the thing: every single one of the tasks that feel so important right now will fade into insignificance if we don't halt climate change. It is our top priority. We need to set aside the distractions, and we need to fight for that golden carbon-zero future. It is possible.

Let's do this.

Resources

A list for readers who want to learn more –

Generation Zero: A grassroots climate action group
developed here in New Zealand
– www.generationzero.org

350 Aotearoa: New Zealand's branch of the global climate
action network
– 350.org.nz

The Crucial Years: Newsletter from Bill McKibben, 350.org
founder
– billmckibben.substack.com

New Zealand Climate Action Network: The local branch of the global CAN network
– nzcan.org/#new-zealand-climate-action-network

Climate Club: A newsletter for anyone who's ever wanted to do something about climate change in Aotearoa New Zealand
– climateclubnz.substack.com

Climate Town: A ragtag team of climate communicators, creatives and comedians here to examine climate change
– youtube.com/c/ClimateTown

Global Weirding: Katherine Hayhoe's lighthearted vlog on climate change
– youtube.com/channel/UCi6RkdaEqgRVKi3AzidF4ow

MarvelClimate: Writing and video from a very eloquent climate scientist
– marvelclimate.com

Climate Adam: Climate change videos from a climate scientist and science communicator
– youtube.com/climateadam

George Monbiot: Insightful commentary on climate science and climate politics
– monbiot.com/category/climate-change

NIWA: New Zealand climate change information
– niwa.co.nz/climate-change

Ministry for the Environment: Facts on how the climate is changing and what action is being taken
– environment.govt.nz/facts-and-science/climate-change

Climate Change Commission: Information and advice on climate action
– climatecommission.govt.nz

Parliamentary Commissioner for the Environment: Independent commentary on Aotearoa's climate and environmental action
– pce.parliament.nz

The Intergovernmental Panel on Climate Change: The authoritative reports on the state of climate change, from 1990 to the present
– www.ipcc.ch